WILDLIFE PEST CONTROL
AROUND GARDENS AND HOMES

by Terrell P. Salmon and Robert E. Lickliter

For information about ordering
this publication, write to:

Publications
Division of Agriculture and Natural Resources
University of California
6701 San Pablo Avenue
Oakland, California 94608-1239

or telephone (415) 642-2431

5m-rep-10/90-MK/HS/PF

Front cover: California ground squirrels, Yolo County.
Jack Kelly Clark

Back cover: California ground squirrel tracks in mud,
Kern County. W. Paul Gorenzel

The authors are Terrell P. Salmon, Wildlife Specialist, and Robert E. Lickliter, Staff Research Associate, Wildlife Extension, Davis. Photographs and art work were compiled by W. Paul Gorenzel, Staff Research Associate, Wildlife Extension.

Acknowledgment

The authors would like to thank the following for their contributions of photographs: Tracy Borland, Jack Kelly Clark, Jerry P. Clark, W. Paul Gorenzel, Carlton S. Koehler, Richard McKillop, Bob Timm, Dale A. Wade, and Heather A. Walker. The art work of Thé Nguyen also deserves appreciation, as does the design and production coordination of Pamela Fabry.

Contents

Introduction

All animals need food, water, and shelter to survive and reproduce. If any of these are in short supply, animal numbers decline proportionately. On the other hand, when these requirements are met, the animal population will reproduce and increase in numbers, remaining stable until the supply of one requirement becomes limiting.

In situations where wildlife species become troublesome or compete with man's economic or health interests, the animals are called pests. Many wildlife species do not thrive in close contact with people. For example, where the human population is dense, the California condor has declined drastically. However, some wildlife species coexist with humans very successfully. In fact, some, like the common rat and mouse, have become more or less dependent on people.

Most wildlife species have the potential of becoming pests: rats and mice may damage food stores or transmit diseases; meadow voles can eat garden crops as well as the bark of young shrubs and trees; pocket gophers can raise mounds of earth that damage farm or garden machinery or result in unsightly lawns and landscape plantings; pigeons may deface buildings; house sparrows may contaminate warehouses or food stores. The most common wildlife pests in California are listed in "quick identification" tables on pp. 2–9. These tables also include a basic description of each pest, its behavioral characteristics, and the kind of damage it is likely to cause.

Wildlife species are only pests in certain situations, usually when their numbers are excessive in a particular area. Human change of an environment will often result in increased numbers of a species. For example, piles of scrap building material make excellent burrowing and hiding sites for rodents, and feeds for livestock and pets are often equally attractive to some wildlife species. In these situations, the pests have suitable food and habitat and will usually multiply rapidly.

An animal population's living situation will affect its size. In many cases we cannot identify the reasons for these adjustments, but factors such as stress, disease, hunger, and fertility probably play a large part. No area can support

(continued p. 10)

1

Quick Identification—Birds

Species	Description
Blackbird 	Many species; 6 to 16 inches; females smaller bodied; sharp, pointed bills; plumage iridescent black; some species have brightly colored areas of yellow, red, or orange on head or wings; female plumage brownish, often with streaked breast.
Crowned Sparrow	Two species, ranging from 5 ¾ to 7 inches; typical sparrow coloration: brownish on back, dull grayish breast. Adult white-crowned sparrow: three white and four black alternating strips on crown. Adult golden-crowned sparrow: dull gold crown margined with black.
Goldfinch	Two species: American Goldfinch, 4 ½ inches; male has bright yellow back and breast, black cap and wings; in winter resembles duller-colored female. Lesser Goldfinch, 3 ¾ inches; male has dark head and back, bright yellow breast; female is dull-colored with dark wings.
Horned Lark 	6 ½ to 7 inches; light brown body above, black band across breast, black strip from bill to eyes; two black "horns" above eyes; walks with slight sideways swaying of body and fore-and-aft movement of head; does not hop on ground.
House Finch	5 to 5 ¾ inches; male has rosy-red head, rump, and breast, brownish back and wings, sides streaked with brown; female lacks red, has brownish body with heavily streaked breast and abdomen.
House Sparrow	5 ¾ to 6 ¼ inches; male has black bib and bill, white cheeks and gray cap; female is dull brown above and dingy whitish below without black bib, bill, or gray cap.

Behavior	Damage
Gregarious; flock ranges from few birds to thousands; some species congregate in huge winter roosts.	Eats vegetables (lettuce, peppers, tomatoes, sweet corn), and nuts (sunflowers, almonds).
Forages on ground in grassy and open areas near brush, fence rows, and other such cover.	Feeds on vegetable and fruit crops, especially lettuce, grapes, melons, almonds, strawberries; disbuds fruit and nut trees; damages young seedlings in fall and winter.
Lives in small flocks in weedy fields, bushes, and roadsides; swoops up and down in flight.	Eats flower and vegetable seeds, strawberries, and sunflowers; disbuds almonds and apricots.
Ground bird found in loose flocks in wide, sparsely vegetated open areas; normally flies low, and swoops up and down slightly.	Feeds on vegetables (lettuce, broccoli, carrots), melons, flowers, particularly at seeding stage.
Well adapted to human environments, often nests in vines on buildings; sings, "chirps" from trees, antennas, or posts; found in variety of habitats, from deserts and open woods to farmlands, suburbs, and farms.	Eats fruits and berries in orchard and garden; disbuds and deflowers fruit and nut trees; attacks seed crops.
Abundant on farms, in cities and suburbs; lives in loose flocks; often nests in buildings in eaves, vents, or other openings and cavities.	Eats emerging seedlings, fruit, buds; damages flowers, newly seeded lawns, ripening fruit; droppings deface buildings.

continued

Birds (continued)

Species	Description

Magpie

Large bird, 16 to 20 inches long; black and white body with long, streaming tail.

Pigeon

14 to 15 inches; plump-bodied, short-billed; usually blue-gray with whitish rump and red feet, but white, brown or other-colored plumage not uncommon.

Scrub Jay

10 to 12 inches; head, wings, and tail blue; underparts and back gray; white throat; no crest.

Starling

7 ½ to 8 ½ inches; short tail; long, slender, yellow bill in spring and summer, dark bill in winter; plumage black to purplish-black; heavily speckled in winter.

Woodpecker (Flicker)

Several species in California, size varies from 5 ¾ to 15 inches; all have strong, sharply-pointed bill for chipping and digging in tree trunks and branches for insects; use stiff tail as a prop. One species, the flicker, is jay-sized woodpecker with brown back, white rump, usually salmon-red under wings, but occasionally yellow.

Behavior	Damage
Lives in farming areas of California valleys and nearby foothills; gregarious, found in colonies; builds large stick nest high in trees near open grasslands or fields.	Feeds on fruits, nuts, grain, garbage.
Found in cities and suburbs; feeds on seeds, grain, fruits, insects; coos intermittently while perched; roosts in large flocks.	Deposits droppings on buildings and cars, contaminates foodstuffs; nests on buildings, may clog drain pipes; transmits disease to humans and domestic animals.
Found throughout California except in deserts and high mountains; very vocal, noisy, makes short flights ending in sweeping glide.	Eats orchard fruits and nuts.
Abundant in city parks, suburbs, and on farms; gregarious; uses large communal roosts from late summer until spring; flies swiftly and directly; primarily ground feeder.	Pulls small plants; damages fruit (grapes, cherries, strawberries, and others); nests in building eaves and other openings; droppings deface buildings.
Most species peck or "drum" repeatedly on resonant limbs, poles, or drainpipes; usually undulate in flight, folding wings against the body after each series of flaps. All species nest in excavated holes. Flickers are often seen on ground eating ants.	Woodpeckers do structural damage, drilling into siding and shingles and under eaves of buildings for food or to excavate nest chamber; damage fences, poles. Drumming on buildings may create annoying noise.

Quick Identification—Mammals

Species	Description
Deer 	Large hoofed mammal, height 3 to 3 ½ feet at shoulder; males have antlers that are shed each year; reddish in summer, blue-gray in winter.
Ground Squirrel	About the size of a tree squirrel, 14 to 20 total inches; tail not bushy; head and body brownish-gray, may have conspicuous dark triangle on back between shoulders.
Meadow Vole 	Chunky body 4 to 5 inches in length; short tail and limbs; small eyes; ears concealed in thick fur.
Mole 	Of four species in California, most common is broad-banded mole; head and body 5 to 6 inches, with pointed snout and 1 ½-inch tail; eyes and ears not visible; soft thick fur, blackish-brown to black; tail slightly haired.

Behavior	Damage
Active mornings and evenings; moves singly or in small groups; follows definite trails.	Feeds on shrub and tree twigs, buds, grasses, and vegetables; males polish their antlers by beating tree and shrub limbs; will invade gardens and orchards, destroying fruits and vegetables.
Lives in colonies in underground burrows; active during daylight hours; hibernates in summer and winter; feeds mostly on ground, but does climb trees; diet depends on season—prefers green vegetation in spring, switches to nut and seed diet in summer.	Damages fruits and nuts such as almonds, apples, apricots, walnuts, oranges, and variety of vegetables; burrows weaken ground above; chews on plastic irrigation pipe; gnaws on trees and shrubs; carries disease.
Lives in burrow and runway system in heavy grass or weed cover (native to hay meadows, irrigated pastures, ditch banks, roadsides, alfalfa fields); seldom found in sparse cover; eats seeds, leaves, and succulent and fleshy vegetation; active day and night, all year round; not found in buildings.	Eats grasses, vegetable seedlings, root crops, and variety of vegetables; girdles trees, shrubs, ornamentals; runways and burrows destroy landscaping.
Rarely comes above ground; lives and feeds in underground tunnels; feeds on insects and earthworms; active day and night; pushes up low ridges as it burrows just under surface; pushes up volcanolike mounds.	Burrows and ridges disfigure gardens, lawns, and landscaped areas. One species, the Townsend's mole, may eat roots and tubers.

continued

Mammals (continued)

Species	Description
Pocket Gopher	Stocky rodent 6 to 12 inches in length; soft fine fur—various shades of brown; small eyes, ears, and flattened head; external cheek pouches and long, curved front claws.
Rabbits 	Two general types in California, both with long, upright ears, long hind legs: Jackrabbits—two species—with very large body, 17 to 22 inches, long ears 5 to 7 inches, short 2- to 3-inch tail. Rabbits—four native species—smaller body and shorter ears, and short, cottony tail.
Roof Rat 	Head and body 7 to 8 inches with tail 8 to 10 inches; naked tail is longer than head and body; large eyes and ears; black or dark brown; light, slender body.
Norway Rat	Rodent with heavy, thick body, 12 to 18 inches in over-all length; tail shorter than head and body; small eyes and ears; blunt snout.
House Mouse	Small rodent, 5 to 7 inches, with tail slightly longer than head and body; light brown to dark gray; almost hairless, scaly tail.
Tree Squirrel	Four to five species in California; depending on species, head and body size varies from 7 to 15 inches, with bushy tail 5 to 14 inches; basic appearance similar to ground squirrel.

Behavior	Damage
Lives and feeds in underground burrow system, seldom seen above ground; pushes dirt from excavations to surface, producing mounds; active all year round; usually one gopher per burrow system.	Mounds cover plants, ruin lawns and landscaping; feeds on roots of trees, shrubs, and garden plants; gnaws on plastic irrigation pipe and underground cables.
Jackrabbits most active early evenings through early mornings. Very fast runners, prefer open spaces, seldom inhabit dense brush or woods. Rabbits (cottontails, brush, and pygmy rabbits) usually found in or near brush or other cover.	Will feed on a variety of garden crops as well as bark, buds, and twigs; girdles small trees, shrubs, vines.
Found chiefly in and around buildings; builds nests in dense vegetation (ivy, shrubs, etc.); in buildings usually found in upper parts (attic or ceiling space between floors); omnivorous, but prefers fresh fruits, nuts, vegetables.	May eat almost any fruit, nut, or vegetable; strips bark and gnaws trees and shrubs; causes structural damage; carries disease; contaminates stored foods.
Found almost anywhere humans are; colonial; lives in burrows; often burrows along foundations of buildings or beneath rubbish piles; omnivorous; nocturnal; uses network of runways.	Will eat variety of fruits, vegetables, and nuts; feeds on anything edible; urine and feces contaminate foodstuffs; carries disease; gnaws; causes structural damage.
Breeds year round; usually found in and around buildings, but also in open fields in burrows; omnivorous.	Feeds on anything edible; damages buildings; destroys and contaminates stored foods; carries disease.
Active during daytime, all year round; rarely ventures far from trees; nests in holes in trees or builds twig and leaf nests in crotch or branches of trees.	Feeds on nuts (green and ripe walnuts, almonds), fruits (oranges, apples, avocados), seeds, and tree cambium beneath bark; gnaws telephone lines and into wooden buildings; invades attics.

(continued from p. 1)
unlimited growth, even if surplus food, water, and shelter are available. Consequently, animals tend to limit their numbers to match the capacity of the area in which they live. If the supply of food, water, or shelter falls below the requirements of the species, its population will decline. This simple concept is the basis for habitat modification as a method of reducing animal numbers in a given area.

Legal Considerations

Most mammals and birds are protected by the California Fish and Game Code or the Federal Code of Regulations. Certain wildlife species that damage growing crops or other property, however, may be controlled in any manner the owner or tenant of the premises chooses. Common home and garden wildlife that are in this category include:

- Norway and roof rats
- House mice
- Pocket gophers
- Ground squirrels
- Meadow voles (meadow mice)
- Jackrabbits
- English sparrows
- Starlings

Special provisions of the California Fish and Game Code must be met if leghold, steel-jawed traps are used. These are rarely recommended, however, for controlling garden pests.

Controlling Wildlife Damage

The primary objective of any control program should be to reduce damage in a practical and environmentally acceptable manner. If you base control methods on a simple knowledge of the habits and biology of the animals causing damage, your efforts will be more effective and will serve to maximize safety to the environment, humans, and other animals.

A key to controlling wildlife damage is prompt and accurate determination of which animal is causing the damage. Even someone with no training or experience can often identify the pest by thoroughly examining the damaged area. Because feeding indications of many wildlife species are similar, other signs, such as droppings, tracks, burrows, nests, or food caches, are usually needed to make a positive species identification. (See photos, pp. 11–26.)

Birds

Starlings

White-crowned sparrow

A brushpile near the garden is a favorable bird habitat.

Birds loafing near garden

11

Bird damage on cherries

Bird damage on nectarines

Bird damage on figs

Bird damage on grapes

Deer

Mule deer

Deer-browsed almond branch

Typical pellet grouping

Unbrowsed 3-year-old lemon

Deer-browsed 3-year-old lemon

Watermelons damaged by deer

Watermelon field damaged by deer

Ground Squirrels

California ground squirrel

Typical burrow surrounded by small pits

Ground squirrel trail in dry grass

Young beet eaten by ground squirrel

Burrow beneath almond tree

A brushpile offers a good ground squirrel habitat.

Mature corn ear completely
stripped by ground squirrel

Branch stripped of almonds by ground squirrels

Strawberries damaged by ground squirrels

Tree bark damaged by ground squirrels

Meadow Voles

Meadow vole

Meadow vole burrows and runways

Grass parted to reveal runway

Meadow vole burrows and runways

Vine trunk girdled by voles

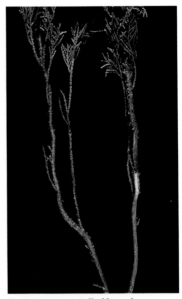

Juniper stems girdled by voles

Cucumber damaged by voles

Vole burrows and runways in lawn

Moles

Mole

Typical volcano-shaped mole hill

Mole hills in lawn

Ridge from burrow near ground surface

Pocket Gophers

Pocket gopher in tunnel system

Typical pocket gopher mound

Pocket gopher feed holes, plugged but with no mounds around them

Grapevine killed by pocket gopher girdling

Pocket gopher mounds in lawn

Below-ground girdling of fruit tree

Rabbits

Cottontail rabbit

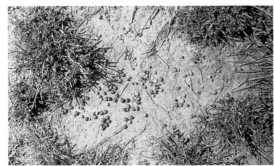

Rabbit droppings

Rabbit droppings

Rabbit-clipped vegetation

Ground cover browsed by rabbits

Rats and Mice

House mouse

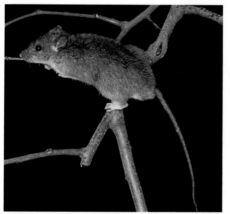

Roof rat

Climbing vegetation is a good habitat for roof rats.

Typical rat harborage

Rat or mouse damage to wall board

Roof rat damage to walnuts

Dead lemon branch killed by roof rat girdling

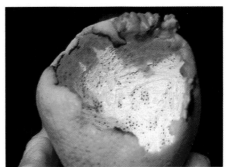

Orange hollowed out by roof rats

Juniper girdled by rats

Tree Squirrels

Western gray squirrel

Fox squirrel at bird feeder

Tree squirrels sometimes enter buildings.

Tree squirrel damage to bark

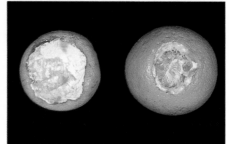

Oranges damaged by tree squirrels

After you properly identify the pest, you can choose control methods appropriate to the animal species involved. Improper control methods may harm but not kill the animal, causing it to become leery of those and other methods in the future. For example, using traps and poison baits improperly or in the wrong situation may teach the animal that the control method is harmful. This may make the animal difficult to control later, even with the correct method.

Four steps lead to a successful wildlife pest control program. You should:

- Correctly identify the species causing the problem.

- Alter the habitat, if possible, to make the area less attractive to the pest.

- Use a control method appropriate to the location, time of year, and other environmental conditions.

- Monitor the site for reinfestation in order to determine if additional control is necessary.

The most successful wildlife control program tailors the technique to the situation, whether it is habitat modification, behavior manipulation, population reduction or a combination of these methods. Table 1 outlines some control methods used on various kinds of mammals causing damage around the home and in gardens. Some of these techniques are discussed on the following pages.

TABLE 1. Most Commonly Used Methods for Controlling Wildlife Pests around Gardens and Homes

Pest	Habitat modification	Repellents	Toxic baits	Glue boards	Traps	Exclusion and protective guards
Deer		•				•
Meadow vole	•		•		•	•
Moles					•	
Pocket gophers			•		•	
Ground squirrels			•		•	
Tree squirrels					•	•
Rabbits		•			•	•
House mice	•		•	•	•	•
Norway rat	•		•	•	•	•
Roof rat	•		•	•	•	•

Detecting the Presence of a Pest

Detecting potential wildlife pests and taking steps to prevent damage is easier, safer, less expensive, and less time consuming than waiting until damage has already occurred. Regular inspection of all buildings, gardens, and surrounding areas will help prevent an increase of wildlife pests. Your best tool to prevent wildlife damage around your home, garden, and property is to know the potential pests and keep your eyes open for them.

Habitat Modification

Modifying an animal's habitat often provides lasting and cost-effective relief from damage caused by wildlife pests. Habitat modification is effective because it limits access to one or more of the requirements for life—food, water, or shelter. Rodent-proofing buildings by sealing cracks and holes prevents potential pests from gaining access to suitable habitats. Storing feed in rodent- and bird-proof containers, controlling weeds and garden debris around homes and buildings, and storing firewood and building supplies on racks or pallets above ground level are also practices that can limit or remove the pests' sources of food, water, or shelter. Habitat modification is not without environmental risks, however, because reducing habitat for the pest species can also reduce it for desirable wildlife such as game and song birds.

Behavior Alteration

Using methods that change the behavior of an animal may lead to a reduction or elimination of damage. Several repellents that are available, such as objectionable-tasting coatings, deter pests from feeding on plants. Other deterrents intended to frighten the animal from the area include bells, horns, and other sound devices, revolving or flashing lights, electrically charged wires, and tacky, sticky substances placed on or near window ledges, or on tree limbs or trunks. Unfortunately, most animals soon discover that repellents and frightening devices are not actually harmful. Once the animals learn this, damage will usually recur.

Population Reduction

Methods such as toxic baits, traps, or shooting may be necessary to reduce pest numbers. But because animals that survive such a program will continue to reproduce as long as they have abundant food, water, and shelter, those methods should be accompanied by sanitation, pest-proof construction and exclusion, and other techniques of habitat alteration whenever possible. When food and habitat are continuously available, as is the case in many gardens, you may need to schedule periodic population reduction control measures.

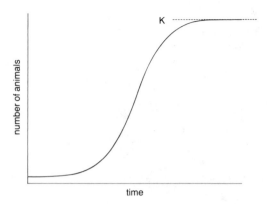

S-shaped population growth curve often observed in animal populations. A small population increases slowly at first; then, under optimum conditions, it increases rapidly before leveling off at K, the carrying capacity.

Natural Control

Effective control programs keep the potential pest population below the damaging level. With all animals, the natural constraints of predators, diseases, and inadequate food and shelter tend to limit populations.

In human environments, these constraints are not as effective. Although predators feed on many small animals in the wild, such predators are not common in most gardens. Even when present, they are usually unable to keep pest animal populations below damaging levels. The introduction of diseases is definitely not recommended for controlling wildlife in gardens or around structures. Reducing sources of available food or shelter is a useful but sometimes impractical preventive measure. Because no population will increase indefinitely, one alternative to a pest problem is to take no action, letting the population limit itself. Experience has shown, however, that by the time the population limits itself, it may already have severely damaged your house or garden.

Wildlife Pest Control Equipment and Supplies

Local retail outlets, such as farm supply and hardware stores, nurseries, and garden shops, often stock pest control supplies. Some county agricultural commmissioners and public health agencies sell or distribute wildlife pest control materials to the public. If appropriate control materials cannot be located, contact your University of California Cooperative Extension Farm Advisor's office—usually listed under *County Offices* in the telephone book—for further information.

Birds

Starling

House finch ♂

House sparrow ♀

House sparrow ♂

Birds are frequent visitors to gardens. Much of the time they pose no threat as they forage for insects and seeds, but during certain periods in a garden's cycle they can cause significant problems. Birds can unearth and eat seeds or feed on newly sprouted seedlings. Occasionally, they will eat buds and flowers, reducing fruit set. (See photos, pp. 11–12.) Probably the greatest damage to the home garden is caused when birds feed on maturing fruits. It often seems that birds have the uncanny ability to start feeding the day before you had planned to harvest.

Many different birds cause problems in garden areas. California's most common bird pests are listed in Table 1, which includes a basic description of each bird, its general behavioral characteristics, and the kind of damage it frequently causes. By accurately identifying the bird causing the problem, you will be able to choose the correct control methods. Some methods are only effective on certain birds. However, most of the techniques discussed here apply to birds found in home gardens.

Legal Restraints of Control

Birds are protected by laws that affect what you can do to solve problems caused by

House finch ♀

White-crowned sparrow

them. Most birds causing problems in gardens are classified as migratory nongame birds according to the U.S. Code of Federal Regulations. Some birds may be taken (controlled) under the general supervision of the County Agricultural Commissioner or under a depredation permit issued by the U.S. Fish and Wildlife Service. Starlings, house sparrows, and pigeons may be controlled without a permit if they are doing damage to your home or garden. If damage is being caused by other birds, contact the local California Fish and Game, U.S. Fish and Wildlife Service, or Agricultural Commissioner's office for information on what to do. However, no permit is needed to scare or exclude depredating birds from the home or garden.

Control Methods

There are several approaches to solving bird problems in gardens. A combination of techniques usually will be the most successful in reducing or eliminating bird problems. Bird damage is often sudden and catastrophic because birds may descend without warning and eliminate your crop in one feeding. Be prepared to take action early!

Observing bird activity in your garden is the first step in preventing damage. Birds, like most animals, establish feeding areas or patterns. Such a pattern may develop into a problem in the future. For example, if your fruit is still green and fruit-eating birds are in the area, you can anticipate that the birds will eat the fruit as it ripens. Therefore, you may want to take some preventive action.

Habitat Modification

Changing the garden to make it less desirable for birds may prevent or reduce damage. Because many birds feed on weed seeds, large quantities of weed seeds in the garden often attract birds to the area. Once the feeding ground has been established, the birds may stay and turn to feeding on crop seeds, seedlings, or fruits and vegetables if they are available.

Brushpiles serve as hiding places for some birds and may cause birds to remain in the garden area. (See photo, p. 11.) By observing the birds in the garden, you may be able to see what kinds of things they like. Do they congregate on brushpiles or clotheslines? If so, try to remove or change these attractions to decrease the desirability of the garden for the birds.

Exclusion

Exclusion is a positive method of reducing or eliminating bird damage around the garden. You can purchase lightweight plastic netting for this purpose. Suspend the netting over seed beds, vines, berries, or small trees in danger of damage from the birds (see fig. I.1). To avoid having the netting reduce or interfere with plant growth, you can suspend it on a trellis or other structure, or place it over the crop just before the crop will be susceptible to damage. Most plastic netting is not damaged by the sun and, with care, should last several seasons. Nets must fit close to the ground around the plants being protected. If gaps occur between the net and the ground, birds can get underneath and damage the "protected" crop. If you are haphazard in placing the net, you will increase the chance of birds becoming entangled in the net, something you want to avoid. Keeping the net taut will help prevent this undesirable situation (see fig. I.2).

Frightening Devices

There are various devices that produce frightening, exploding booms, high-pitched sounds, or electronic noises. Recordings of alarmed birds may be played to frighten birds. Although such devices are effective in some situations, most are loud and will disturb people in the area. For this reason, they are not recommended for gar-

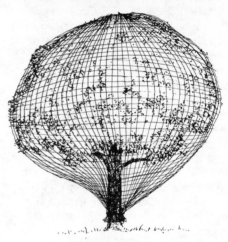

FIGURE I.1. Two examples of bird netting over fruit trees

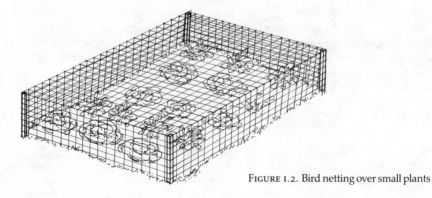

FIGURE I.2. Bird netting over small plants

den use. Visual frightening devices, such as scarecrows, dangling pie tins, moving shapes, artificial owls or snakes, and flashing lights, are rarely if ever effective. The effectiveness of all frightening methods diminishes greatly as birds become accustomed to them.

Trapping

Starlings, house sparrows, and pigeons can be trapped without a permit. Trapping other species requires a permit issued by the U.S. Fish and Wildlife Service or authorization by the local County Agricultural Commissioner. Trapping can prevent or reduce bird damage, particularly if the birds live in or near the garden area. Most trapping programs are quite selective. Plan the trapping to fit the existing garden conditions and the species to be trapped. Each species varies greatly in its vulnerability to being caught with different traps and different trap baits. Trapping is not usually effective on birds that migrate seasonally into your area because they can arrive in large numbers in a very short period of time.

Traps can be homemade or purchased through many hardware and farm supply stores. Trap types vary according to the kind of bird to be trapped. Figure I.3 depicts an automatic trap design for house sparrows. Figure I.4 depicts a double funnel trap, also for house sparrows.

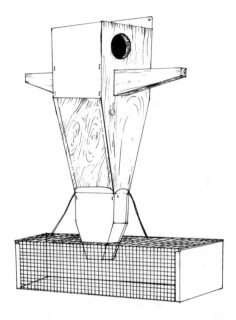

FIGURE I.3. An automatic multicatch house sparrow trap. Upon entering upper compartment, birds drop into the wire cage, from which there is no escape.

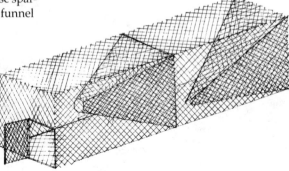

FIGURE I.4. Double funnel trap for house sparrows. Birds are attracted by baits scattered in front of and inside trap. They enter apex of funnel forming entrance. Birds then go through opening of apex of second funnel, above floor, into second compartment, from which escape is almost impossible.

To place the trap effectively, observe where the birds are flying into the garden, resting, perching, and feeding to determine the best place to set the trap. Traps are usually most effective when placed in the open along an entry route and a short distance from a perching or feeding site.

An effective bait may be food that the birds are already feeding on or some other food item. If you are using bait and it does not seem to be working, change to something else. Place the bait inside the trap, with a small amount outside the trap entrance to act as enticement. With many traps, leaving a few live birds inside will serve to attract others. The decoy birds must be supplied with food, water, perches, and shelter from cold winds or hot sun.

Disposing of birds. Most traps are live-catch types, so they present the problem of disposing of the live bird. Releasing birds nearby generally negates the effects of trapping because the birds will simply return. Also, most birds trapped once are reluctant to enter the trap again; others learn to enter the trap for food. Birds may be killed by im-mersing the trap in water or placing it in an enclosed box and fumigating the box with exhaust from an automobile.

Other Control Methods

Toxic baits and repellents control bird damage in some agricultural situations, but they are not recommended for garden-type problems.

Repellents are used to keep birds from nesting and roosting on buildings; however, these have little, if any, application for preventing bird damage in gardens.

Monitoring Guidelines

Closely watching your garden is the best method of monitoring potential bird problems. Learn to identify the birds and know what damage they may cause. If the birds that are present threaten your crop, develop a plan of action you can initiate when the birds arrive so damage can be minimized.

Deer

Black-tailed deer

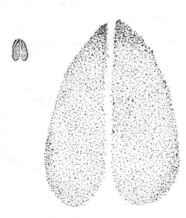

Black-tailed deer footprint, actual size *(above)* , and typical pattern of tracks *(left)*

Many people enjoy deer. Unfortunately, deer can be very destructive to gardens, orchards, and landscaped areas, particularly in foothill and coastal districts where nearby woodlands provide deer with cover. Deer may damage a variety of plants, including vegetables, fruit and nut trees, grape and berry vines, grasses, and many ornamentals. They cause damage by eating as well as trampling crops. (See photos, pp. 13–14.) Young trees or shrubs may also be damaged when deer rub their antlers on trunks and limbs.

Mule deer (*Odocoileus hemionus*) and black-tailed deer (*O. hemionus columbianus*) are the two species common in California. These deer eat a variety of vegetation, including woody plants, as well as some grasses and forbs (small broad-leaved flowering plants). They also consume fruits, nuts, ornamental trees, shrubs, vines, and garden vegetables. Because most deer feed in the late evening and very early morning, it is not easy to observe them. A good way to determine their presence in the garden or orchard is to

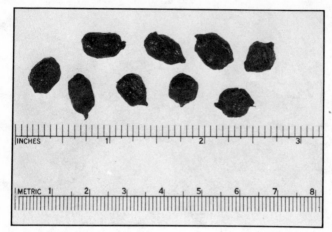

Deer pellets, actual size

look for hoofprints. Deer hooves are split, pointed at the front and more rounded at the rear, and are about 2 to 3 inches long.

Legal Restraints of Control

Deer are classified as game animals by the California Fish and Game Code. If you find them damaging property or crops, you may obtain a permit from your local game warden to control deer by shooting them, although this method is not generally recommended for the problems homeowners encounter. Other methods of destroying deer, such as the use of traps, poisons, or toxic baits, are illegal. Deterrents such as fences, barriers, and repellents can all be used without a permit.

Control Methods

Physical exclusion and, to a lesser extent, repellents are recommended for protecting gardens, orchards, and ornamental plantings from deer. In cases where such methods are not practical, contact your local farm advisor for further information.

Fencing

Properly built and maintained fencing is the most effective method for excluding deer. Deer normally will not jump a 6-foot fence for food, but, if threatened, can jump an 8-foot fence on level ground. While 6-foot upright fences are usually adequate on level ground, a 7- or 8-foot fence is recommended, especially in the Sierra Nevada areas of California where larger deer are found. On sloping ground, you may need to build fences 10 or 11 feet high to guard against deer jumping from above.

Determine the kind of fence you are going to build by assessing your needs, expense, and terrain. Woven mesh wire attached the full height of the fence is preferable. (See fig. II.1.) If you need to economize, two or more strands of 9- or 10-gauge smooth wire can be stretched at 4- to 6-inch spacings above a 5-foot mesh. Vertical stays should not be more than 6 to 8 inches apart. There is no advantage in using barbed wire. Because deer will crawl under a fence if they can, you should secure mesh wire close to ground level. An extra strand of barbed wire stretched along the ground will help prevent deer from crawling under. Stake the wire firmly to the ground in any depressions between posts, or fill the depressions with materials that will not deteriorate or wash away.

Gates. With upright fences, gate height should be approximately equal to fence height. Keep weight to a minimum. A light wooden frame over which mesh wire is stretched is often satisfactory. If you use factory-made aluminum gates, you may bolt or weld on metal extensions and stretch mesh wire over them. It is advisable to sink a metal or treated wooden base frame in the ground below the gate to make a uniform sill and to prevent deer from working their way under the gate.

Slanting fences. In general, upright fences have proved most satisfactory. Under some conditions, however, a slanting fence is cheaper to construct and is advantageous because of its lower height. The slanting fence is effective because it acts psychologically as a barrier to deer. Deer usually first try to crawl under such a fence. When they find this impossible and see the wire extended above them, they are discouraged from jumping. Slanting fences are effective

in one direction only: slanting away from the area to be protected.

Overhanging or slanting fences are particularly suitable for temporary fencing because fewer posts are used and the wire can be more easily rolled when it is no longer needed. Slanting fences are also suitable for locations where an upright fence would be unsightly or otherwise unsuitable. The principal disadvantages of overhanging or slanting fences are that cattle and horses can easily damage them, hogs are likely to push under them, and weeds growing underneath may render them ineffective.

Overhanging or slanting fences are not recommended when it is necessary to use minimum horizontal space. In heavy snowfall areas, mesh wire slanted at a 45° angle or less is apt to be crushed by the settling snow pack; smooth wires stretched horizontally at 4-inch spacings are more useful in such circumstances.

FIGURE II.1. A deer fence showing two methods of using wire—either mesh alone or a combination of mesh and smooth wire.

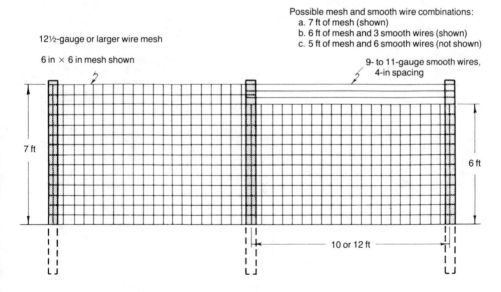

Possible mesh and smooth wire combinations:
a. 7 ft of mesh (shown)
b. 6 ft of mesh and 3 smooth wires (shown)
c. 5 ft of mesh and 6 smooth wires (not shown)

12½-gauge or larger wire mesh

6 in × 6 in mesh shown

9- to 11-gauge smooth wires, 4-in spacing

7 ft

6 ft

10 or 12 ft

Converting existing fences. To convert upright fences with wooden posts (see fig. II.2), scab 2- by 4-inch boards onto the post tops either horizontally, vertically, or at an angle. Fasten mesh wire or smooth wires with no more than 4- to 6-inch vertical spacing to the two-by-fours. Extensions of steel posts must be bolted or welded on. Vertical extensions, whether on steel or wooden posts, are the easiest to attach, and an upright extension of 2 or 3 feet on most mesh wire fences will make them deer-proof, provided the lower portion is well constructed.

Electric fences. Experience has shown that electric fences of the standard designs used for livestock have usually proved unsatis-factory for deer control in California. How-ever, future designs in electric fencing may be better.

Maintaining fences. Deer fences of any design must be checked regularly. Make necessary repairs on damaged wire, broken gates, soil washout beneath fences, or any weakness in construction that would permit deer access. The job becomes increasingly difficult as fences age and become more vulnerable to breakage.

Other Control Methods

In many places, protecting individual plants may be more practical and economical than attempting to exclude deer from an

FIGURE II.2. Three ways of converting a common upright fence to a deer fence.

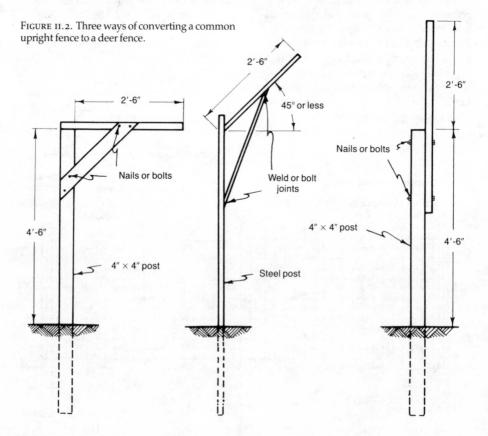

entire area (see fig. II.3). For example, young fruit or nut trees in a home orchard can be individually fenced until primary branches grow above the reach of deer. Two or more wooden stakes can be driven into the ground, and chicken wire or heavier woven wire can be attached to form a circle around the tree. Plastic trunk protectors may be useful for young vines and trees. Inspect individual barriers regularly.

Various chemical repellents are available as a means of reducing or preventing deer damage to trees, vines, and ornamentals. Deer repellents are distasteful materials that make the protected plants less desirable as food sources for deer. It is important to remember that repellent materials must be noninjurious to the trees or shrubs. Also, do not apply repellents to edible crops unless such use is specifically indicated on the product label.

Repellents are useful under some conditions. Most are not registered for use on food crops except during the plants' dormant season. They should be tested to make sure they are not phytotoxic (harmful to the plant). When deer are hungry and a garden area contains highly preferred foods, repellents probably will not be effective. Repellents are ineffective with dense, severely competitive deer populations as well.

When you use deer repellents, follow label directions carefully. Most repellents should be applied before damage occurs and must be reapplied frequently, especially after a rain, heavy dew, or sprinkler irrigation. Likewise, to be effective, repellents must be applied to new foliage as it develops. Some repellents produce odors thought to frighten or repel deer from an area. Examples are human hair balls and lion dung or other types of predator excrement. Although these substances may have some temporary repelling effect, they have not

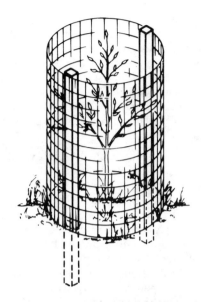

FIGURE II.3. A 4- to 5-foot wire cage to protect a plant from deer browsing. One- or 2-inch mesh is best. Diameter will vary with plant size.

proved satisfactory in reducing deer damage in California.

Noisemaking devices, such as propane cannons and electronic alarms, are not effective generally in repelling deer from home or garden areas because deer rapidly adjust to them. In addition, noise from such devices can disturb neighbors.

Modifying an area to eliminate suitable shelter or the other survival needs of deer is rarely possible. Garden and landscape trees, shrubs, and vines are often favored foods, especially when new foliage is forming on them. Planting other foliage that is an attractive food source to distract deer near the home or garden probably will not prevent damage to more valued plants and might even make the area as a whole more attractive to deer.

Yet deer, like all animals, have certain food aversions. Home gardeners living near a deer habitat can often take advantage of this fact by using deer-resistant plants for ornamental planting. Various factors can make a plant resistant to deer. Many of the most resistant plants (such as oleander) are poisonous, some at all times and others only at certain stages of growth. Palatability of nontoxic plants also varies with plant age and time of year.

Resistance of the plant to deer is also related to the availability of other food. If there is an adequate supply of native-plant food, ornamental plantings may be largely untouched. If the naturally occurring plant-food supply is low, there will be increased browsing in domestic gardens. If there is an extreme shortage of natural food, few plant species will be totally resistant to deer. A heavy deer population also increases competition for food, with the result that plants normally unpalatable to deer may be browsed.

Table 2 lists some common deer-resistant plants. This table should be used only as a general guide. It was taken from the University of California Division of Agriculture and Natural Resources Leaflet 2167 entitled "Deer-Resistant Plants for Ornamental Use." Only plants in the leaflet that are classified "particularly resistant" are listed. For a more complete list, refer to the publication.

Monitoring Guidelines

The first step to preventing damage from deer is to know if they are in your garden. Although the animals are large and easily seen, their nocturnal feeding habits necessitate examining the garden at night with a flashlight. Look for physical signs of deer such as tracks, droppings, trails, and damage to foliage from deer feeding. Because a few deer can do a lot of damage to a garden or landscaped area, take action when deer signs are first detected. If you know deer have caused problems nearby, consider using exclusion methods such as fences before damage occurs.

TABLE 2. Botanical (and Common) Names of Plants That Show Some Resistance to Deer Browsing

Adolphia californica
Agapanthus africanus (blue lily-of-the-Nile)
Agave spp. (century plant)
Aloe spp. (aloe)
Aquilegia spp. (columbine)
Aralia spinosa (Hercules club)
Artemisia tridentata (basin sagebrush)
Arundo donax (giant reed)

Beaucarnea recurvata
Buddleia davidii (butterfly bush)
Buxus spp. (boxwood)

Cactaceae spp. (cactus)
Calycanthus occidentalis (western spice bush)
Casuarina stricta (she oak)
Chamaerops humilis (European fan palm)
Choisya ternata (Mexican orange)
Clematis spp. (clematis)
Correa spp. (Australian fuchsia)
Cotinus coggygria (smoke tree)
Cytisus scoparius (Scotch broom)

Daphne spp. (daphne)
Datura spp.
Delphinium spp. (larkspur)
Digitalis spp. (foxglove)
Diospyros virginiana (persimmon)

Echinocystis lobata (wild cucumber)
Echium fastuosum (pride of Madeira)
Erythea armata (Mexican blue palm)

Fabiana imbricata (Chile heath)
Ficus spp. (fig)
Fraxinus velutina (Arizona ash)
Furcraea spp.

Gelsemium sempervirens (Carolina jessamine)

Hakea suaveolens (sweet hakea)
Hedera helix (English ivy)
Helleborus spp. (hellebore)
Hippophae rhamnoides (sea buckthorn)

Ilex spp. (except thornless) (holly)
Iris spp. (iris)

Jasminum spp. (jasmine)

Kerria japonica (Japanese rose)
Kniphofia uvaria (devil's poker, red-hot poker)

Leptodactylon californicum (prickly phlox)
Leucojum spp. (snowflake)
Lupinus spp. (lupine)
Lyonothamnus floribundus (Catalina iron-wood)

Melaleuca leucadendra (cajeput tree)
Melia azedarach (China-berry tree)
Melianthus major (honey bush)
Mesembryanthemum spp. (ice plant)
Myrica californica (wax myrtle)

Narcissus spp. (narcissus, daffodil, jonquil)
Nerium oleander (oleander)
Nolina parryi (nolina)

Oxalis oregana (oxalis, redwood sorrel)

Phoenix spp. (date palm)
Phormium tenax (New Zealand flax)
Prunus caroliniana (Carolina cherry laurel)
Pueraria thunbergiana (kudzu vine)

Quillaja saponaria (soapbark tree)

Rhododendron spp., except azalea-leaved varieties (rhododendron)
Rhus ovata (sugar bush)
Robinia pseudoacacia (black locust)
Romneya coulteri (matilija poppy)
Rosmarinus officinalis (rosemary)

Sabal blackburniana (hispaniolan palmetto)
Sambucus racemosa (red elderberry)
Schinus molle (California pepper tree)
Schinus polygamus (tree pepper)
Solanum spp. (nightshade)
Spartium junceum (Spanish broom)
Syzygium paniculatum (Australian brush-cherry, eugenia)

Tecomaria capenisis (cape honeysuckle)
Teucrium fruticans (germander)
Trachycarpus fortunei (windmill palm)
Trillium spp. (trillium, wake robin)
Tulipa spp. (tulip)

Washingtonia spp. (Washington palm)

Zantedeschia spp. (calla lily)
Zauschneria spp. (zauschneria, California fuchsia)

Ground Squirrels

California ground squirrel

California ground squirrel footprints, actual size (*above*), and a typical pattern of tracks (*left*)

Ground squirrels inhabit most agricultural and rural areas of California. They are found around buildings, gardens, and industrial sites, as well as in nature. This section provides basic information about controlling ground squirrels. However, when their numbers are high or they are found over large acreages, other control methods may be more practical or economical.

Ground squirrels damage many food-bearing and ornamental plants. Particularly vulnerable are all types of grains, and fruits and nuts such as almonds, apples, apricots, peaches, pistachios, prunes, oranges, tomatoes, and walnuts. Ground squirrels eat

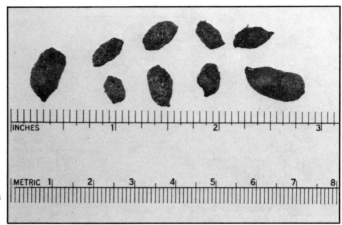

Ground squirrel droppings, actual size. Often found in small pits near burrow opening.

certain vegetables and field crops at seedling stage. Squirrels damage young shrubs, vines, and trees by gnawing bark, girdling trunks, eating twigs and leaves, and burrowing around roots. Squirrels even gnaw surface-type plastic irrigation pipe. (See pp. 15–17.)

While burrowing, ground squirrels can also be quite destructive. In the process of digging burrows, ground squirrels make large mounds of soil and rock that may bury and kill grass or other small plants. Burrows and mounds make it difficult to mow and harvest, and present hazards to machinery. Ground squirrels frequently burrow around trees and shrubs, damaging the root systems and sometimes killing the plants. Burrows beneath buildings and other man-made structures sometimes necessitate repair or replacement.

Ground squirrels can transmit diseases (such as tularemia and plague) to humans, particularly when squirrel populations are dense. Do not handle dead squirrels. If you notice unusual numbers of dead squirrels or other rodents, notify public health officials.

There are two main species and several subspecies of ground squirrels in the state. The

California ground squirrel (*Spermophilus beecheyi*) is found in most areas except the Mojave Desert. The Belding or Oregon ground squirrel (*S. beldingi*) is found at higher elevations in the northeast counties and regions of the Sierra Nevada mountains.

The two species can be distinguished without much difficulty; the California ground squirrel is the larger. Its head and body measure 9 to 11 inches and it has a somewhat bushy tail, 5 to 9 inches in length. The Belding ground squirrel's head and body measure 8 to 9 inches. It has a relatively short tail, 2 ½ to 3 inches long.

Ground squirrels live in a wide variety of natural habitats, but populations may be particularly dense in areas disturbed by humans such as road or ditch banks, fence rows, around buildings, and in or bordering many crops. They usually avoid thick chaparral, dense woods, and wet areas. They live in colonies of 2 to 20 or more animals. Much of their time is spent underground in burrows where they sleep, rest, rear young, store some food, and escape danger.

Ground squirrels are active during the day and are easily seen, especially in warm

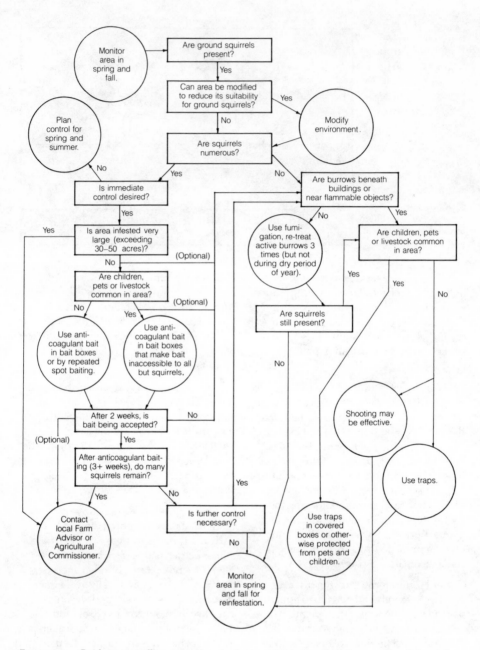

FIGURE III.1. Guide to controlling ground squirrels around structures, gardens, and small farms.*

*Additional factors, such as costs and practicality of various options, must also be taken into account.

44

weather from spring to fall. During winter months, most ground squirrels hibernate, but it is common for some young to remain active, especially in areas where winters are not severe. Most adults go into a summer hibernation (estivation) during the hottest times of the year.

Ground squirrels reproduce once a year in early spring. Litter sizes vary, but seven to eight young are average. The young remain in the burrow about 6 weeks before they emerge.

Ground squirrels are primarily vegetarians. During early spring, they consume green vegetation such as grasses and forbs. When the vegetation begins to dry, squirrels eat seeds, grains, and nuts, and begin to store food. Squirrels eat fruits and vegetables and are known to eat or gnaw bark from bushes and trees.

Legal Restraints of Control

Ground squirrels are classified as nongame mammals by the California Fish and Game Code. If you find ground squirrels to be injuring growing crops or other property of which you are the owner or tenant, you may take (control) the squirrels in any manner. You must satisfy special provisions of the California Fish and Game Code if you wish to use leg-hold, steel-jawed traps.

Ground squirrels should not be confused with tree squirrels, which are classified as game animals. Although ground squirrels can climb trees, the two species can usually be distinguished by their response to danger. Frightened ground squirrels retreat to a burrow; tree squirrels climb a tree or high structure.

Control Methods

When ground squirrels cause severe damage, a control program of procedures suitable for the situation and time of year can alleviate the problem. Figure III.1 is a guide to such an approach. Figure III.2 depicts the activity cycle of ground squirrels and when control measures are appropriate. A program incorporating these procedures should result in significant reductions in ground squirrel populations in the area.

Trapping

Traps are practical devices for controlling ground squirrels in small areas where the

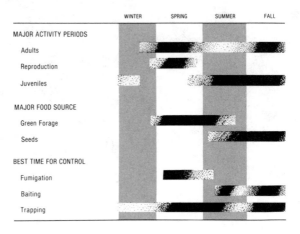

FIGURE III.2. Many control methods are effective against ground squirrels only at certain times of the year. This chart shows the yearly activities of the California ground squirrel and times when baiting, trapping, and fumigating should be carried out.

45

number of squirrels is moderate. Live-catch traps are effective, but present the problems of how to kill the live squirrels and where to dispose of them. Because ground squirrels carry diseases and are agricultural pests, it is illegal to release them elsewhere.

There are several types of traps that kill ground squirrels. Most types work best if you place them on the ground near squirrel burrows or runways. Walnuts, almonds, oats, barley, and melon rinds are attractive trap baits. Place bait well behind the trigger or tied to it. Bait the traps but do not set them for several days so the squirrels become accustomed to them. After the squirrels are used to taking the bait, rebait and set the traps.

A box-type squirrel trap (fig. III.3) kills ground squirrels quickly. This trap is available commercially or can be constructed from a box-type wooden gopher trap. To modify a gopher trap, lengthen the trigger slot with a rat-tail file or pocket knife so the trigger can swing unhindered and the squirrel can pass beneath the swinging loop of the unset trap. Remove the back of the trap and replace it with hardware cloth, which allows the animal to see the bait from both ends but prevents it from entering the trap from the back.

For a dual-assembly trap, place two box traps back-to-back and secure them to a board (fig. III.4). A small strip of hardware cloth connects them and forms the baiting

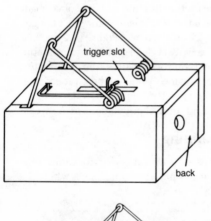

trigger slot

back

FIGURE III.3. Box-type squirrel trap

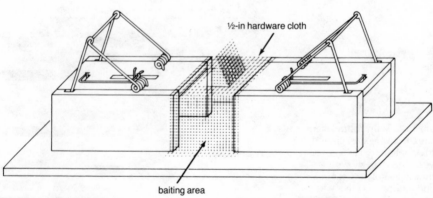

½-in hardware cloth

baiting area

FIGURE III.4. Multiple-catch box-type ground squirrel trap

FIGURE III.5. Conibear trap set in burrow entrance. Secure trap with a stake.

area. Place the bait through a small door cut through the wire or through the open ends of the trap.

To construct a multiple-trap box, place box traps inside a lidded box. Traps can be placed in the entrances or inside the box. When placed inside, the traps are assembled side by side and a narrow rear baiting area is formed by a perpendicular baffle. Enclosed traps minimize the chances that you will accidentally catch pets or poultry.

The Conibear trap is also an effective ground squirrel kill trap. The wire trigger permits the trap to be used either baited or, more commonly, without bait. The trap is placed so the squirrel will pass through it, tripping the trigger. It is best to set the Conibear trap directly in the burrow opening or where physical restrictions in the squirrel runway will direct the animal through the trap (fig. III.5). Do not place it where pets or other nontarget animals are likely to pass. Placing traps in a covered box will reduce hazard to children, pets, and poultry. When you are using a Conibear trap, leaving the trap baited but unset has little effect on trapping success except when traps are inside a box. In that case, bait should be

used and the traps left unset until squirrels enter the box freely.

Other ground squirrel traps are available in some areas. With all traps, take precautions to reduce the hazard of trapping nontarget wildlife, pets, and poultry.

Fumigation

Ground squirrels can be killed in their burrows by several types of toxic gases, some of which require a special permit from the County Agricultural Commissioner. Fumigation should not be used beneath buildings. It is most effective in the spring or at other times when soil moisture is high. At those times, gas is contained within the burrow system and does not diffuse into small cracks which are often present in dry soil.

Ground squirrel burrows are quite large and can have several entrances. Treat all entrances and then seal them. Fumigation is not effective during periods of hibernation or estivation because the squirrel plugs its burrow with soil. The plug is not obvious to a person examining the burrow entrance.

The U.S. Fish and Wildlife Service produces a relatively safe and easy-to-use gas

a

FIGURE III.6. Sequence for fumigation with the Fish and Wildlife Service gas cartridge: *a)* Use nail, screwdriver, or sharp instrument to punch holes in one end of gas cartridge and insert fuse into one of holes; *b)* place gas cartridge into burrow and light fuse;

b

c

c) once you are sure the contents have ignited, use a shovel handle to push the gas cartridge further into burrow; *d)* immediately close burrow with soil and pack it tightly to prevent escape of gas.

d

cartridge designed for fumigating burrowing rodents. (See fig. III.6.) The cartridge is a mixture of chemicals which, when ignited, gives off suffocating gas. It is available at many County Agricultural Commissioners' offices and can be used without special permit. Several commercially manufactured fumigation cartridges are available at retail outlets. When using any cartridge-type fumigant, follow package instructions.

Use a cartridge when fresh digging indicates the presence of an active ground squirrel burrow. With a nail or similar sharp object, puncture the cap end of the cartridge at the points marked, rotating the nail to loosen the material inside. Insert the fuse into the center hole. Place the cartridge in the burrow as far back as possible and light the fuse. With a shovel handle or stick, push the lighted cartridge down the burrow and quickly seal the opening with soil, tamping it down lightly. Seal connected burrows if smoke is seen escaping. Larger burrow systems usually require two or more cartridges placed in the same or connecting burrow openings. After 24 hours, re-treat reopened burrows.

Gases emitted from the cartridge occasionally ignite, creating some fire danger. Therefore, do not use gas cartridges where a significant fire hazard exists, such as under buildings or near dry grass or other flammable material.

Toxic Baits (Rodenticides)

Toxic baits have long been used for controlling ground squirrels. Some toxic baits are available over the counter and others require a permit for their use that the County Agricultural Commissioner issues. In general, baits requiring a permit are not used by gardeners. When you use toxic bait or any other rodent-control materials, *follow label instructions carefully*.

Anticoagulant baits are recommended for controlling ground squirrels because they are effective against the pest and relatively safe to humans and pets. Anticoagulants interfere with an animal's blood-clotting mechanisms, eventually leading to death. They are effective only when consumed in several feedings over a period of 5 or more days. Effectiveness is greatly reduced if 48 hours or more elapse between feedings. These features, as well as an effective antidote (vitamin K_1), make the use of anticoagulant baits relatively safe.

Anticoagulant baits can be used in two ways: in bait boxes or by repeated spot baiting. Bait boxes are small structures that the squirrel must enter to eat the bait. (See fig. III.7.) Boxes contain sufficient bait for re-

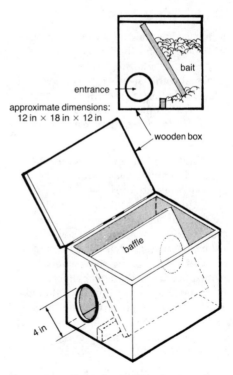

FIGURE III.7. Example of bait box for ground squirrels: *a)* A simple ground squirrel bait station used for anticoagulant baits.

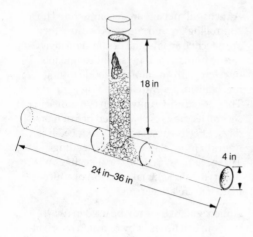

18 in

4 in

24 in–36 in

FIGURE III.7. Example of bait box for ground squirrels: *b*) A PVC-pipe anticoagulant bait station for ground squirrels.

peated feedings and help keep children and pets from reaching the bait. Bait boxes are the preferred baiting method around homes and other areas where children, pets, and poultry are present. Unless a bait label specifies otherwise, bait boxes can be constructed from any durable material and in a variety of designs.

There are several things to consider when you are designing a bait box for ground squirrels. The entrance hole(s) should be about 4 inches across to allow access to squirrels but not to larger animals. Construct a lip to prevent bait from spilling out of the box when squirrels exit. Provide a lock on the box or devise some other method that will make it difficult for children to open the box. The bait box should be secured so it cannot be turned over or easily removed. A self-feeding arrangement will insure that the pest gets a continuous supply of bait.

Place bait boxes containing 1 to 5 pounds of bait in areas frequented by ground squirrels (near runways or burrows, for example).

(See fig. III.8.) If ground squirrels are noticeable throughout the area, space the boxes at intervals of 100 to 200 feet. Initially, inspect bait stations daily and add bait as needed. Increase the amount of bait if all is eaten overnight. Fresh bait is important; replace moldy or old bait. It may take a number of days before squirrels become accustomed to the bait box and enter it. Anticoagulant bait generally requires 2 to 4 weeks or more to be effective. It does not immediately affect feeding habits of squirrels. Continue baiting until all feeding ceases and no squirrels are observed. You should pick up and dispose of unused bait upon completion of the control program.

Repeated spot baiting (without a bait box) with anticoagulant bait can be effective in controlling ground squirrels. Follow label instructions. If spot or broadcast baiting is not specified on the product label, *do not use that baiting method*.

Anticoagulant baits have the same effect on nearly all warm-blooded animals, including birds. Cereal baits are attractive to some dogs as well as to other nontarget animals, so take care to prevent their access to the bait. Danger to children and pets can be reduced by placing bait out of their reach, as in a bait box. Dead ground squirrels should be buried or discarded in plastic bags. Do not handle them with your bare hands. In case a person or pet ingests anticoagulant bait, contact a physician or veterinarian immediately.

Natural Control

As with all animals, natural constraints such as inadequate food and shelter, predators, disease, and bad weather limit ground squirrel populations. Experience has shown, however, that in most environments altered by humans, the point at which squirrel populations level off naturally is intolerably high.

FIGURE III.8. Bait box spacing around a garden. Concentrate boxes where squirrel activity is highest.

Ground squirrels generally are found in open areas, although they usually need some cover to survive. Removing brushpiles and debris not only makes an area less desirable to ground squirrels, but also makes detection of squirrels and their burrows easier, aids in monitoring the population, and improves access during control operations.

Many predators—including hawks, eagles, rattlesnakes, gopher snakes, and coyotes—eat ground squirrels. In most cases, predators are not able to keep ground squirrel populations below the level at which they become pests. Predators sometimes can prevent ground squirrels from invading marginal habitats where cover is not abundant. Dogs may prevent squirrels from en-

tering small areas, but they cannot control squirrel populations.

Monitoring Guidelines

Once ground squirrel damage has been controlled, a system should be established to monitor the area for squirrel reinfestation. Observe from an isolated structure or automobile during the morning hours when squirrels are most active. Ground squirrels may move in from other areas and cause new damage within a short time. Experience has shown that it is easier, less expensive, and less time consuming to control a population before it builds up to the point where damage is excessive.

Meadow Voles

Meadow vole

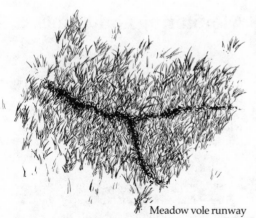

Meadow vole footprints, actual size

Meadow vole runway

Meadow voles (*Microtus californicus*), also known as meadow or field mice, damage a wide range of plants by feeding and gnawing on trunks, roots, stems, leaves, and seeds. They are common in hay fields, irrigated pastures, and some row crops. They also invade gardens, orchards, and landscaped areas.

Meadow voles are small rodents with heavy bodies, short legs and tails, and small, rounded ears. Their long, coarse fur is blackish brown to grayish brown in color. When full grown, they are 4 to 5 inches long.

Meadow voles are active all year long and are normally found in areas with dense ground cover. They are poor climbers and do not usually enter buildings. Instead, they are found in gardens or other vegetated sites. (See pp. 18–19.)

Meadow voles dig short, shallow burrows and make underground nests of grass, stems, and leaves. The peak breeding period is spring with a second, smaller breeding period in fall. Litters average four young. Meadow vole numbers fluctuate from year to year; under favorable condi-

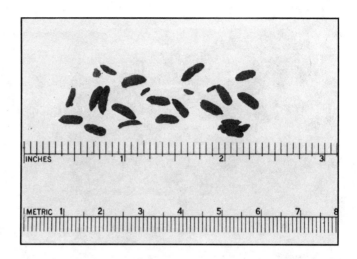

Meadow vole scat, actual size

tions, their populations increase rapidly. It is during such times that most problems around structures and gardens occur.

Legal Restraints of Control

Meadow voles are classified as nongame mammals by the California Fish and Game Code. Nongame mammals injuring or threatening growing crops or other property may be taken (controlled) at any time and in any manner by the owner or tenant of the premises.

Control Methods

Preventing meadow vole damage usually requires a management program that keeps down the population in the area. This can often be achieved by removing or reducing the vegetative cover, making the area unsuitable to voles. (See fig. IV.1.) Removing cover also makes detecting voles and other rodents easier. When necessary, a program for reducing the vole population should be undertaken. Because the damage these animals do to ornamental and garden plants

can be quite severe, and because of their rapid reproductive rate, initiating a program of habitat modification or population reduction before their numbers explode is most desirable.

Habitat Modification

Habitat modification is particularly effective in deterring voles. Weeds, heavy mulch, and dense vegetative cover encourage meadow voles by providing food and protection from predators and environmental stresses. If you remove their protection, the area will be much less suitable to the voles. Clearing grassy areas adjacent to gardens can be helpful in preventing damage because it will reduce the base area from which voles invade gardens or landscaped areas. Weed-free strips can also serve as buffers around areas to be protected. The wider the cleared strip, the less apt meadow voles will be to cross and become established in gardens. A minimum width of 15 feet is recommended, but even that can be ineffective when vole numbers are high. Buffer strips are most useful around young trees or vines, or in other areas where the voles will have to remain in the open to feed.

a

b

FIGURE IV.1. *a)* Weeds around trunks offer excellent habitat for meadow voles; *b)* Meadow vole habitat is reduced with good weed control.

Wire or metal barriers at least 12 inches high with a mesh size of ¼ inch or less will exclude meadow voles from garden areas. (See fig. IV.2.) Meadow voles rarely climb such fences, but they may dig beneath them. To prevent digging, bury the bottom edge 6 to 10 inches. Young trees, vines, and ornamentals can also be protected by individual hardware cloth cylinders that surround their trunks. (See fig. IV.3.) Such devices must be supported so they cannot be pushed over or pressed against the trunk. The bottom should be buried below the soil surface to prevent the voles from digging under them.

Repellents

Several commercial repellents are registered and available for protecting plants from meadow voles. Apply them before damage occurs. Voles usually damage plants at or just beneath the soil surface, making adequate coverage difficult. Because repellents are often washed off by rain, sprinklers, or even heavy dew, you must re-apply them to give continued protection to the garden. Repellents should not be applied to food crops unless such use is specified on the product label.

Trapping

When voles are not numerous or when the population is concentrated in a small area, trapping may be an effective control method. The simple, wooden mouse trap is commonly used. (See fig. IV.4.) Peanut butter, oatmeal, or apple slices make excellent baits for meadow voles. Often, no bait is

FIGURE IV.2. Small-mesh wire attached to bottom of a fence will exclude meadow voles.

FIGURE IV.3. Trunk protector for meadow voles

needed because voles will trigger the trap as they pass over it.

Trap placement is crucial. Meadow voles seldom stray from their usual travel routes, so set traps along these routes. Look for nests, burrows, and runways in grass or mulch in or near the garden. Place baited traps at *right angles* to and flush with the ground in these runways.

Traps must be set in sufficient numbers to be effective in controlling the population. A dozen traps for a small garden is probably the minimum number required, and 50 or more may be needed for larger areas. Examine traps daily. Remove and bury dead voles or place them in plastic bags in the trash. Do not handle voles without rubber gloves.

Toxic Bait

When meadow voles are numerous or when damage occurs over large areas you may need to use toxic bait to achieve adequate control. When you use toxic baits, take care to ensure the safety of children, pets, and nontarget animals. Follow product label instructions carefully.

Multiple-feeding baits. Anticoagulant baits are slow acting and must be consumed over a period of 5 or more days to be effective. They are therefore probably the safest type of rodent bait for use around homes and gardens. Many types and brands of anticoagulant baits are available. Whole grain baits are commonly recommended, but pelleted baits work too. Moisture-resistant paraffin block baits are useful

FIGURE IV.4. Set snap trap perpendicular to meadow vole runway.

around ditches and other areas where high moisture may cause other types of baits to spoil.

Because the pest must feed on most antico-agulant baits over a period of 5 days if they are to be effective, the bait must be available until the vole population is controlled. As with trapping, bait placement is very important. Place it in runways or next to burrows so voles will find it during their normal travel. (See fig. IV.5.) Usually, baiting every other day for 5 days will be effective. Be sure to broadcast (spread) the bait evenly over an infested area if that application method is specified on the label. If you use this technique, you will probably have to broadcast every other day for three or four treatments.

Paraffin bait blocks can also be used. Place them in runways or near burrow openings or both. Keep replacing them as they are eaten and remove those that remain when feeding stops. Bait blocks should not be used where children or pets might pick them up.

Single-feeding baits. Baits that require only one feeding to be lethal are called single-feeding baits. They are particularly useful where vole populations are spread over large areas. Zinc phosphide is the most common single-feeding bait used for meadow voles around structures. Place bait in runways or next to burrows where voles will find it. A problem with this method is "bait shyness," a condition that results when voles eat only enough bait to make them sick. If this happens, the voles will not eat zinc phosphide bait again for 6 months or more. To prevent this, use the bait according to label directions and do not treat with zinc phosphide bait more often than every 6 months.

Zinc phosphide bait is rapid acting. You may find dead voles within 12 hours of baiting. Dispose of all dead voles by burying them or placing them in plastic bags and putting them in the trash. Do not handle them with your bare hands. Because zinc phosphide does not accumulate in the tissue of the voles, predators or scavengers such as dogs and cats are not likely to be ad-

FIGURE IV.5. Place bait in meadow vole runway or next to burrow opening.

versely affected by eating the poisoned rodents. However, children, as well as pets and other animals, can be affected by the bait, so store it out of reach and use it carefully in a way that will minimize their access to it.

Zinc phosphide is a Restricted Use Material, and you must obtain a permit from the local County Agricultural Commissioner to buy and use it. Homeowners may be exempt from the permit requirement in some situations, such as when small amounts of bait are purchased.

Natural Control

As with all animals, natural population constraints limit meadow vole numbers. Because populations will not increase indefinitely, one alternative is to do nothing, letting the voles limit themselves. Experience

has shown, however, that around homes and gardens, the natural population peak is too high and damage will still occur.

Predators, especially raptorial birds, eat meadow voles. However, in most cases, predators are unable to keep vole populations below damaging levels.

Monitoring Guidelines

To detect the presence of voles, look for fresh trails in the grass, as well as burrows, droppings, and evidence of feeding. Routine monitoring of the garden and surrounding area is important. Pay particular attention to adjacent areas that have heavy vegetation because such areas encourage invasions.

Moles

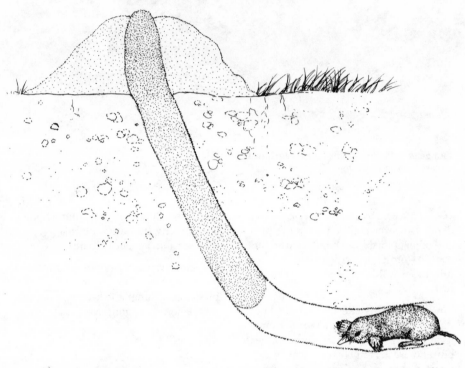

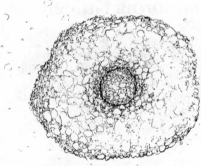

Mole mounds tend to be circular with a plug in the middle. Compare with pocket gopher mound.

The mole (*Scapanum* spp. and *Neurotrichus gibbsii*) is a small, insect-eating mammal, not a rodent as many people think. In California, moles inhabit the Sierra and Coast Range mountains and foothills, as well as the entire coastal zone. They are not generally found in the Central Valley or dry southeastern areas of the state. Moles live almost entirely underground in a vast network of interconnecting tunnels 3 to 30 inches deep. They feed mainly on worms, insects, and other invertebrates but also eat some roots, bulbs, and seeds. Their burrowing sometimes dislodges plants, and their mounds and ridges are often annoying to gardeners.

Moles have cylindrical bodies with slender pointed snouts, and short and bare or sparsely haired tails. Their limbs are short and spadelike. Their eyes are poorly developed and their ears are not visible. The fur is short, dense, and velvety. Moles have one litter of three or four young during early spring.

The mounds formed by moles are pushed up from an open center hole. The soil may be in chunks, and single mounds often appear in a line over the runway connecting them. Main runways are usually less than 2 inches in diameter, and may be 16 to 18 inches below the surface. Surface feeding burrows appear as ridges that the mole pushes up by forcing, not digging, its way through the soil just below ground level. (See p. 20.) Moles are active throughout the year, although surface activity slows during periods of extreme cold or drought.

Legal Restraints of Control

Moles are classified as nongame mammals by the California Fish and Game Code. Nongame mammals that are found to be injuring growing crops or other property may be taken (controlled) at any time and in any manner by the owner or tenant of the premises.

Control Methods

Moles can cause considerable problems in landscape or garden areas. Various methods are available, and you may need to use a combination of techniques to control the pests adequately.

Trapping

Trapping is the most universally applicable and dependable method of mole control. A number of different mole traps are available at hardware stores or nurseries, or directly from the factory.

An understanding of mole behavior helps improve your trap set. When a mole's sensitive snout encounters something strange in the burrow, the mole is likely to plug off that portion and dig around or under the object. For this reason, traps are generally set straddling or encircling the runway, or are suspended above it. Most mole traps operate on the theory that a mole will push its way through a soil block in its tunnel. Setting a mole trap successfully relies on the fact that moles are not suspicious of fine soil blocks in their runway because cave-ins happen naturally. The mole will readily push its way through the soil block to reopen the tunnel. The trap is sprung by pressure from the mole's body or the movement of soil against a triggering plate.

Moles are active throughout the year and can be trapped at any time. Before setting mole traps, determine which runways are in current use. Moles dig a system of deep tunnels as well as a network of surface runs. Some of the surface tunnels are only temporary runs dug in the search for food. These may not be reused, whereas the deep runways are more or less permanently used.

To determine where moles are active, stamp down short sections of surface runways. Observe these areas daily and restamp any raised sections, remembering the areas of activity. The selection of a frequently used runway is very important to the success of your control efforts. Set traps at least 18 inches from a mound, and only in those runways used daily by the mole. You can locate deeper tunnels by probing between, or next to, a fresh mole hill with a

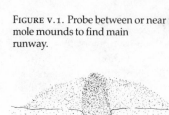

FIGURE V.1. Probe between or near mole mounds to find main runway.

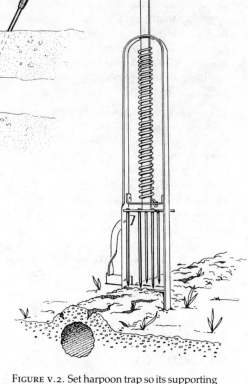

pointed stick, slender metal rod, or standard gopher probe (see fig. V.1). When the earth suddenly gives way to your probe, the probe has probably broken through the burrow.

Mole traps are fairly expensive, so most people tend to buy only one. Although one trap may solve the problem, increasing the number of traps will often drastically increase the speed and overall success of the trapping program. In areas where moles are common, stores handling mole traps, as well as tool rental outlets, may rent traps on a daily or weekly basis.

In California, several types of mole traps are available. The Reddick (harpoon type, fig. V.2) and the Out-O-Sight (scissor-jaw type, fig. V.3) mole traps are the most often used. Moles have also been caught with Macabee gopher traps set in mole runways. Trap manufacturers often provide detailed instructions which should be followed carefully.

The harpoon trap will work in the deeper tunnels if you set it on a dirt plug as described for the scissor-jaw trap. It can also be set on the surface over an active runway ridge that has been pressed down under the trigger pan.

FIGURE V.2. Set harpoon trap so its supporting stakes straddle runway. This trap may also be set in main runway like scissor-jaw trap in figure V.3.

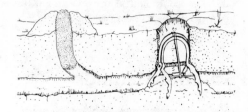

FIGURE V.3. Set scissor-jaw trap so that it straddles runway. Remember to fill the portion of the tunnel under trap's trigger with loose soil.

Set the scissor-jaw trap in the mole's main underground tunnel, usually 6 to 10 inches deep. Using a garden trowel or small shovel, remove a section of soil slightly larger than the trap width (about 6 inches). Build a plug of soil in the center of the opened runway for the trigger pan to rest on. (Moist soil from the opened tunnel or from a nearby fresh mound can be pinched together to build the plug.) Wedge the set trap, with safety catch in place, firmly into the opened burrow with the trigger placed snugly against the top of the dirt plug. Take care that the opened trap jaws do not protrude into the open ends of the mole's tunnel lest the animal become suspicious. Now sift loose dirt onto the set trap to about the level of the coil spring to exclude light from the opened burrow and make the mole less suspicious of the plugged tunnel. Release the safety catch and the trap is completely set.

Toxic Baits

Because the mole's main diet is earthworms and insects, poisoning with the strychnine grain baits currently registered in California is rarely effective. However, where a mole population covers large areas, as in some agricultural situations, people use poison baits. Even though they are not highly effective, poison baits are more economical than traps. If you choose this technique, you will probably need to follow it with trapping.

Other Control Methods

Some gardeners have found that mole movements can be detected by watching the ridges caused by the mole's surface runs. If you can see such movements, try using a shovel or pitchfork to kill the mole.

People who have tried flooding an area and fumigating it with various gases and chemicals have not had great success. Such methods are therefore not recommended.

If moles are deprived of their food supply, they will move to other areas to find food. Several insecticides are capable of reducing earthworms and insects in the soil. An insect control program may indirectly result in reduced mole populations.

A number of other methods have been suggested to solve mole problems. These include placing irritating materials such as broken glass, razor blades, rose branches, bleach, moth balls, lye, and even human hair in the burrow, or using frightening devices such as mole-wheels, vibrating windmills, and whistling bottles. Another reported method is the use of repelling plants such as gopher or mole plant (*Euphorbia lathyris*). None of these approaches has proved successful in stopping mole damage or in driving moles from an area.

Monitoring Guidelines

Once you have controlled damage, establish a system to monitor for reinfestation. Mounds and surface runways are easy to detect. These are indicators of reinfestation. Because mole damage is frequently unsightly, the number of moles that can be tolerated is usually quite low, sometimes even none at all. As soon as you see an active mound or surface runway, you should initiate appropriate control actions.

Pocket Gophers

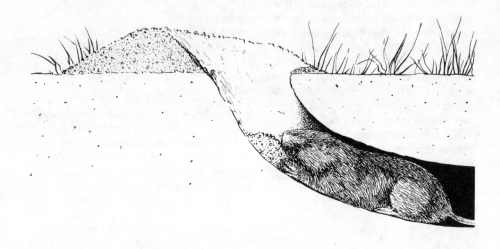

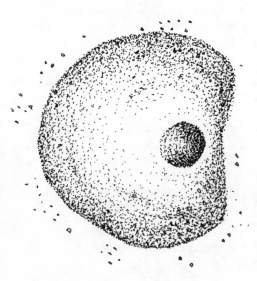

Typical pocket gopher mound, cross-section and from above. The pocket gopher pushes soil out *(top)*, creating a fan-shaped mound *(above)*. Gopher then closes hole with a soil plug. Compare with mole mound.

Pocket gophers (*Thomomys* spp.) are stout-bodied, short-legged rodents, well adapted for burrowing. They live by themselves in an extensive underground burrow system that can cover an area of several hundred square feet. These burrows are about 2 inches in diameter, usually located from 6 to 12 inches below ground. Pocket gophers often invade gardens and lawns and feed on many agricultural crops. They eat a wide variety of roots, bulbs, tubers, grasses, and seeds, and sometimes even the bark of trees. Their feeding and burrowing can cause damage to lawns, ornamental plants, vegetables, forbs, vines, and trees. (See pp. 21–22.) In addition, they may damage plastic water lines and lawn sprinkler systems; their tunnels can divert and carry off irrigation water and lead to soil erosion.

Pocket gophers range in length from 6 to 12 inches. They have a thick body with little evidence of a neck, and their eyes and ears are quite small. They have a good sense of smell, which they use to locate their food. Their common name is derived from the fur-lined external cheek pouches, or pockets, used to carry food and nesting materials. The pocket gopher's lips can be closed behind its four large incisor teeth, keeping dirt out of its mouth when it is using its teeth for digging.

Pocket gophers seldom travel aboveground. They are sometimes seen feeding, pushing dirt out of their burrow system, or moving to a new area. The mounds of fresh soil that are the result of burrow excavation indicate their presence. Such mounds are usually crescent shaped and are located at the ends of short lateral tunnels branching from a main burrow system. One gopher may create several mounds in a day.

Legal Restraints of Control

Pocket gophers are classified as nongame mammals by the California Fish and Game Code. Nongame mammals that are found to be injuring growing crops or other property may be taken (controlled) at any time and in any manner by the owner or tenant of the premises.

Control Methods

Because of the nature of pocket gopher damage, a successful control program depends on early detection and promptly applied control measures appropriate to the location and situation. Most people control gophers in lawns, gardens, or small or-

chards by trapping them or by using poison baits placed by hand. A program incorporating these methods should result in significant reduction in pocket gopher damage in the area.

Successful trapping or baiting depends on accurately locating the gopher's main burrow which runs in both directions at a depth between 6 and 18 inches. The crescent-shaped mounds visible aboveground are connected to this burrow by lateral tunnels. Because the lateral tunnels are plugged by the gopher, trapping and baiting in them is not as successful.

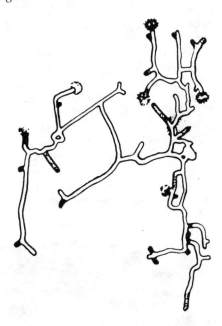

FIGURE VI.1. Aerial view *(above)* and cross-section *(below)* of typical pocket gopher burrow systems.

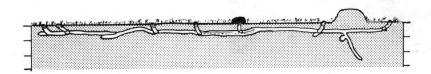

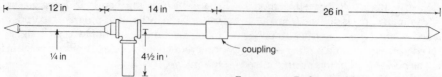

FIGURE VI.2. Probes used for locating pocket gopher tunnels can be built at home. The shaft may be in one piece or divided by a pipe coupling for convenient carrying when not in use.

To locate the main burrow, use a gopher probe. Gopher probes are commercially available or can be constructed from a pipe, wooden dowel, or stick. An example of a gopher probe is given in figure VI.2. Look for the freshest mounds, because they indicate an area of recent gopher activity. You will usually see a small circle or depression representing the plugged lateral tunnel. This plug is generally bordered on one side by soil, making the mound form a crescent shape. Begin probing 8 to 12 inches from the plug side of the mound. When the probe penetrates the gopher's burrow, it should drop suddenly about 2 inches. Often, the main burrow will go between two mounds. To locate the gopher's main burrow you will probably have to probe repeatedly, but your skill will improve with experience.

Trapping

Trapping can be a safe and effective method to control pocket gophers. Several types and brands of gopher traps are available. The most commonly used is a two-pronged pincher trap (also called the Macabee trap, see fig. VI.3) which is triggered when the gopher pushes against a flat vertical pan. Another popular trap is the squeeze-type box trap.

After you have located the main tunnel, open it with a shovel or garden trowel and set traps in pairs facing opposite directions. This placement will intercept a gopher com-

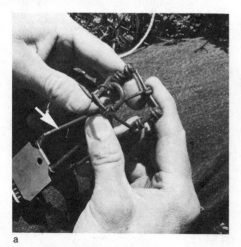

FIGURE VI.3. To set the Macabee trap: *a)* Hold trap exactly as in illustration; be sure your index finger holds trigger *(arrow)* up through trap frame; *b)* press your thumbs down to spread sharp jaws; use your index finger to guide trigger hook *(arrow)* over end of trap frame; *c)* still holding frame down, place other end of trigger into small hole *(arrow)* in plate; *d)* place set trap, sharp jaws first, into gopher burrow;

a

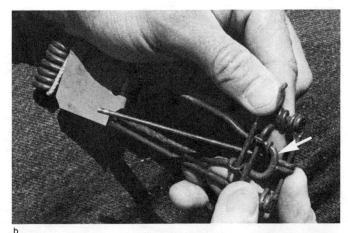

b

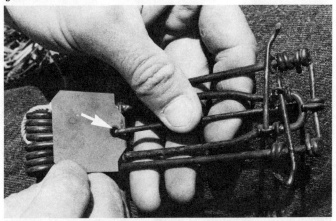

c

d

e

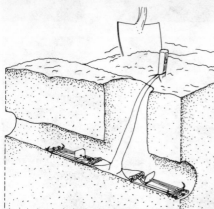

f

e), f) place a second set trap in burrow, facing in opposite direction; fasten each trap by wire or cord to a stake; push traps well into burrow; *g)* cover hole to block any light from entering burrow.

g

ing from either end of the burrow. The box type is easier for most inexperienced trappers to set but requires more excavation, an important consideration in lawns and some gardens. Box traps are useful when the diameter of the gopher's main burrow is small (less than 3 inches) because small burrows must be enlarged to accommodate wire traps. (See fig. VI.4.) All traps should be wired to stakes so you won't lose track of them. After setting the traps, exclude light from the burrow by covering the opening with dirt clods, sod, cardboard, or some other material. Fine soil can be sifted around the edges to ensure a light-tight seal. If light enters, the gopher may plug the burrow with soil, filling the traps and making them ineffective. Check traps often and reset them when necessary. If no gopher is caught within 3 days, reset the traps in a different location.

Baiting

Strychnine-treated bait is the most common type used for pocket gopher control. This bait generally contains 0.25 percent to 0.5 percent strychnine and is effective with one application. Baits containing anticoagulants are available in some areas. They require multiple treatments or one large treatment to be effective. Gopher bait is poisonous and should be used with caution. Read and follow product label instructions carefully.

Always place pocket gopher baits in the underground tunnel. After locating the main gopher burrow with a probe (see fig. VI.5), enlarge the opening by rotating the probe or inserting a larger rod or stick. Then place the bait carefully in the opening, taking care not to spill any on the ground surface (see fig. VI.6). (A funnel is useful for preventing spillage.) Close the probe hole with sod, rock, or some other material to exclude light and prevent dirt from falling on the bait. Tamp down existing mounds so you can distinguish new activity. If gopher mound-

FIGURE VI.4. Box trap set in lateral runway. A pair of traps should be used when set in main runway.

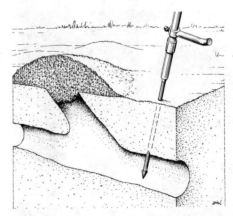

FIGURE VI.5. Use probe for placing pocket gopher baits. When probe suddenly drops about 2 inches, a main tunnel has been located. Enlarge probe hole enough to insert poisoned bait.

ing activity continues for more than 2 days after strychnine baiting or 7 to 10 days after anticoagulant baits have been used, you will need to use further control methods.

Exclusion

Some protection may be achieved by using ½-inch mesh fence buried 2 feet and extended aboveground 2 feet. However, because gophers burrow extensively, this expensive method is less than perfect. Small areas such as bulb beds may be protected from pocket gophers by complete underground screening with ½-inch mesh wire. If you use wire, be careful to put the wire deep enough so it will not restrict root growth.

Other Control Methods

Pocket gophers can easily withstand normal garden or home landscape irrigation, but flooding can sometimes be used to force them out of their burrows where they become vulnerable to predation, usually human. Fumigation with smoke or gas cartridges is usually not effective because gophers quickly seal off their burrow when they detect smoke or gas.

No repellents currently available will successfully protect gardens or other plantings from pocket gophers. The plant gopher purge (*Euphorbia lathyris*) has been suggested as a repellent but no conclusive evidence exists as to its effectiveness. Frighten-

ing gophers with sounds, vibrations, electromagnetic radiation, or other means has not proved effective.

Because no population will increase indefinitely, one alternative to a gopher problem is to do nothing, letting the population limit itself. Experience has shown, however, that by the time gopher populations level off naturally, much damage has already been done around homes and gardens.

Predators, especially owls, eat pocket gophers, but in most cases they are unable to keep pocket gopher populations below the levels that cause problems in gardens and landscaped areas.

Monitoring Guidelines

Once pocket gopher damage has been controlled, a system should be established to monitor the area for gopher reinfestation. Level all existing mounds after the control program and clean away weeds and garden debris so fresh mounds can be seen easily. A monitoring program is important because pocket gophers may move in from other areas and a recurrence of damage can occur within a short time. Experience has shown that it is easier, less expensive, and less time consuming to control gophers before they build up to the point where they do excessive damage.

FIGURE VI.6. To probe and hand-bait for pocket gophers: *a)* Use probe to find gopher burrow; *b)* after you feel a noticeable give, use shaft of probe to enlarge hole; *c)* insert a funnel into hole and slowly pour bait down funnel into burrow; *d)* remove funnel and place a clod of dirt over hole to block any light from entering tunnel.

Rabbits

Cottontail rabbit

Rabbits are a form of wildlife enjoyed by many people. Unfortunately, they can be very destructive in gardens and landscaped areas. Rabbits eat a wide variety of plants, including grasses, grains, alfalfa, vegetables, fruit trees, vines, and many ornamentals. (See p. 23.) They also damage plastic irrigation systems.

There are three species of rabbits common to California: the jackrabbit (*Lepus californicus*); the cottontail (*Sylvilagus audubonii*), and the brush rabbit (*S. bachmani*). Because of its greater size and abundance, the jackrabbit is the most destructive of the three in California.

The jackrabbit is about as large as a house cat. It has long ears, short front legs, and long hind legs. Jackrabbits typically occupy open or semiopen lands in California valleys and foothills. They do not build a nest but make a depression in the soil beneath a bush or other vegetation for seclusion. When born, young jackrabbits are fully

Cottontail rabbit footprints, actual size *(above)*, and a typical pattern of tracks *(left)*.

Black-tailed jackrabbit

haired and their eyes are open. Within a few days, they can move about quite rapidly.

Cottontail and brush rabbits are smaller and have shorter ears. They generally inhabit places with dense cover such as brushy areas, wooded areas with some underbrush, or areas with piles of rocks or debris. Their young are born naked and blind and remain in the nest for several weeks.

Legal Restraints of Control

Jackrabbits are classified as game mammals by the California Fish and Game Code. If they injure growing crops or other property, they may be taken (controlled) at any

Black-tailed jackrabbit footprints, actual size *(above)*, and a typical pattern of tracks *(right)*.

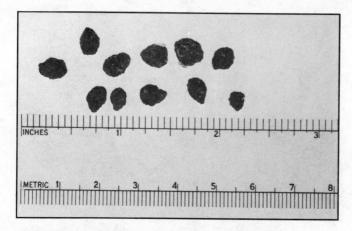

Cottontail rabbit pellets, actual size.

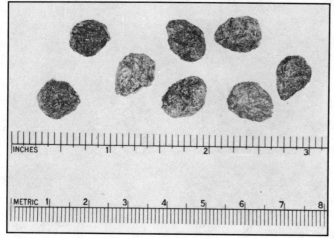

Black-tailed jackrabbit pellets, actual size

time or in any manner by the owner or tenant of the premises. Cottontail or brush rabbits are also game mammals, and may be taken by the owner or tenant of the land, or by any person authorized in writing by such owner or tenant, when the rabbits are damaging crops or forage. Any person other than the owner or tenant of the land must be carrying written authority of the owner or tenant of the land where the rabbits were taken at the time rabbits are being transported from the property. These rabbits cannot be sold.

Control Methods

A number of methods can be used to reduce rabbit damage but physical exclusion, trapping, and, to a lesser degree, repellents are recommended for protecting garden and home areas. In cases where these

72

FIGURE VII.1. Rabbit fences must be attached securely into ground. Burying base of wire will help prevent rabbits from digging under.

methods are not practical, contact your local Farm Advisor or Agricultural Commissioner for further information.

Exclusion

Fences, if properly built, can be very effective in keeping rabbits out of an area. A 30- to 36-inch-high wire fence with a mesh no larger than 1 inch, with the bottom turned outward and buried 6 inches in the ground, will exclude rabbits. (See fig. VII.1.) Include tight-fitting gates with sills to keep rabbits from digging below the bottom rails. Keep gates closed as much as possible because rabbits can be active day or night. Inspect the fence regularly to make sure rabbits or other animals have not dug under. Poultry netting supported by light stakes is adequate for rabbit control, but larger animals, especially livestock, can damage it easily. Cottontail and brush rabbits will not jump a 2-foot fence. Jackrabbits ordinarily will not jump a fence this high unless they are being chased by dogs or otherwise frightened. Discourage jumping by increasing the aboveground height to 3 feet. Several strands of smooth or barbed wire strung above the wire mesh may also be used to increase the height of the fences. Remember, once a rabbit gets into the fenced area, it may not be able to get out.

In many places, protecting individual plants may be more practical than attempting to exclude rabbits from the entire area. (See fig. VII.2.) Poultry netting with a 1-inch mesh, 18 to 24 inches wide, can be cut into strips 12 to 18 inches long and formed into cylinders to be placed around the trunks of young trees and shrubs. Bury the bottom of the cylinders 2 to 3 inches and brace them away from the trunk so rabbits cannot press against the trees or foliage and nibble through the mesh. As you would with a protective fence, inspect these barriers regularly. Commercial tree trunk protectors are also available.

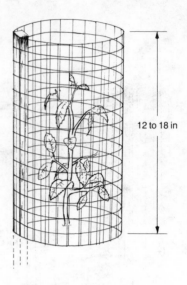

12 to 18 in

FIGURE VII.2. Individual plant protector for rabbits can be made with 1- to 2-inch mesh wire. Fence should be 12 to 18 inches high, with bottom buried in 2 to 3 inches of soil.

Repellents

Various chemical repellents can reduce or prevent rabbit damage to trees, vines, or ornamentals. Rabbit repellents make the protected plants less desirable to eat by making them taste bad. Such repellents must be noninjurious to trees or plants.

Repellents are useful under limited conditions. Some are not safe for use on plants or plant parts to be eaten by humans. When rabbits are hungry, or the garden area contains highly preferred foods, repellents probably will not be effective. Most repellents should be applied before damage occurs and must be reapplied frequently, especially after a rain, heavy dew, or sprinkler irrigation, or when new growth occurs. In all cases, follow the label directions for the repellent you are using.

Trapping

Trapping is not effective for jackrabbits because of their reluctance to enter a trap. Trapping with a box or similar type trap is often effective for cottontail and brush rabbits if their numbers are not large. Another simple way to trap rabbits is to construct a small corral along a rabbit-tight fence surrounding the protected area. Construct a short strip of fence at a diagonal to the main fence, funneling the rabbits through a one-way gate into the corral. Inspect the corral daily. Because rabbits can carry certain diseases and are considered agricultural pests, it is illegal to release them in other areas.

Other Control Methods

Guns or dogs can be an effective means of eliminating small numbers of rabbits. Best results are achieved in early morning or evening when rabbits are most active.

FIGURE VII.3. Live traps can be used to catch cottontail rabbits.

Check local regulations for any restrictions on shooting in your area.

Predators, especially hawks and coyotes, eat rabbits. However, in most cases, these predators are unable to keep rabbit populations below damaging levels.

Monitoring Guidelines

Rabbits are large and easily seen but, because they frequently feed during darkness, you may have to examine the garden at night with a flashlight to see them. Additionally, look for signs of rabbits such as droppings, trails, and feeding damage. Generally, if rabbits are feeding in an area, droppings can be found nearby. Because few if any rabbits can be tolerated in a garden or landscaped area, take appropriate action when signs of rabbits are first observed. Rabbits that have been observed nearby will frequently invade the garden when the plantings become desirable to them. Therefore, consider exclusion methods or possibly an area-wide control program before damage actually occurs.

Rats and Mice

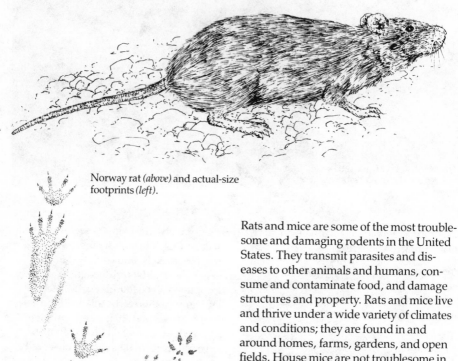

Norway rat *(above)* and actual-size footprints *(left)*.

House mouse *(below)* and actual-size footprints *(right)*.

Rats and mice are some of the most troublesome and damaging rodents in the United States. They transmit parasites and diseases to other animals and humans, consume and contaminate food, and damage structures and property. Rats and mice live and thrive under a wide variety of climates and conditions; they are found in and around homes, farms, gardens, and open fields. House mice are not troublesome in garden and landscaped areas; however, they are likely to invade nonmouseproof buildings. Rats can cause significant damage to garden crops and ornamental plantings. (See pp. 24–25.)

House mice (*Mus musculus*) are small, agile rodents. They have slender bodies and weigh ½ to 1 ounce. House mice have rather large ears for their body size and their semi-hairless tail is about as long as the body and head combined. Mice will eat a wide variety of foods, although they seem to prefer cereal grains. They have keen senses of taste, hearing, smell, and touch and are excellent climbers. In addition, mice can squeeze through openings as small as ¼ inch. They generally nest in secluded places such as crates, boxes, sacks, drawers, and

Roof rat *(above)* and actual-size footprints *(right)*.

walls. In a single year, a female may have five to ten litters, averaging five young each. House mice are sometimes seen darting from one secluded place to another during the day or night. Other habits also provide numerous signs of their presence: sounds, droppings, urine stains, smudges and gnawing marks, footprints, nests, and musky mouse odors.

The Norway rat (*Rattus norvegicus*), sometimes called the brown or sewer rat, is much larger than a house mouse, averaging 7 to 10 ounces. Its burrows are found along building foundations, beneath rubbish or woodpiles, and in and around gardens and fields. When Norway rats invade buildings, they usually remain in the basement or ground floor. They are most active at night and eat a wide variety of foods. They have keen senses of hearing, taste, and smell, and can swim and climb quite well. They can enter buildings through any opening larger than ½ inch.

Roof rats (*R. rattus*), sometimes called black rats, are slightly smaller than Norway rats. Their tail is longer than their head and body. This trait can be used to distinguish them from Norway rats, whose tail is shorter than their head and body combined. Roof rats are agile climbers and usually live and nest in shrubs, trees, and dense ground cover such as ivy. In buildings, they are most often found in enclosed spaces in attics, walls, and cabinets. Roof rats are most active at night and, although they eat a wide variety of foods, they prefer fruits, nuts, and berries.

People do not often see Norway and roof rats, but their habits provide numerous signs of their presence. Such signs include squeaking and scratching noises, droppings, urine stains, smudges and gnawing marks, feet and tail tracks, nests, food caches, and damaged fruits, vegetables, or household goods.

Legal Restraints of Control

Rats and mice are classified as nongame mammals by the California Fish and Game Code. Nongame mammals that are found injuring growing crops or other property may be taken (controlled) at any time and in any manner by the owner or tenant of the premises.

Control Methods

To successfully control rats and mice you must work on three fronts:

- Sanitation measures
- Building construction and rodent-proofing
- Population control

Good sanitation is fundamental to effective rodent control but is less successful with house mice than with rats. Sanitation will reduce high populations of house mice but mice can survive on relatively small amounts of food. Neat, off-the-ground storage of lumber, firewood, crates, boxes, sacks, gardening equipment, and other household items will help reduce the suitability of the area for mice and will also make their detection easier.

Good housekeeping in and around buildings will reduce available shelter and food sources for Norway and to some extent roof rats. (See fig. VIII.1.) Neat, off-the-ground storage of pipes, lumber, firewood, crates, boxes, gardening equipment, and other household goods will help reduce the suitability of the area for rats and will also make their detection easier. Garbage, trash, and garden debris should be collected frequently, and all garbage receptacles should have tight-fitting covers.

For roof rats in particular, thinning dense vegetation will make the habitat less desirable. Climbing hedges such as Algerian or English ivy growing on fences or buildings are very conducive to roof rat infestations and should be removed if possible. Separate the canopy of densely growing plants such as pyracantha and juniper from each other and from buildings by a distance of one foot or more to prevent movement between them by rats. Sanitation is fundamental to rat control and must be a continuous effort or the benefits of other measures will be lost and rats will quickly return.

The most successful form—and a permanent one—of rat and mouse control is to "build them out." Seal cracks and openings

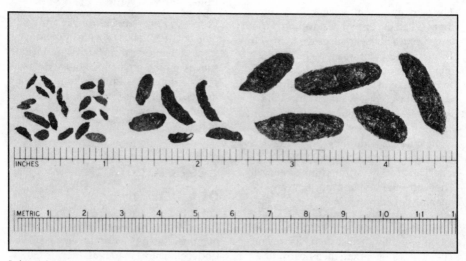

Left to right: House mouse, roof rat, and Norway rat droppings, actual size.

in building foundations, and any openings for water pipes, electric wires, sewer pipes, drain spouts, and vents. No hole larger than ¼ inch should be left unsealed. Make sure doors, windows, and screens fit tightly. Their edges can be covered with sheet metal if gnawing is a problem. Coarse steel wool, wire screen, and lightweight sheet metal are excellent materials for plugging gaps and holes. Plastic sheeting, wood, or other less sturdy materials are likely to be gnawed away. Because rats and mice are excellent climbers, openings above ground level must also be plugged.

When food, water, and shelter are available, rat and mice populations can reproduce impressively. While the most permanent form of control is to limit food, water, shelter, or access to buildings, direct population control is often necessary.

Trapping

Under most conditions, trapping is the safest and easiest method for controlling rats and mice in and around homes, garages, and other structures when only a few animals are present. The simple wooden snap

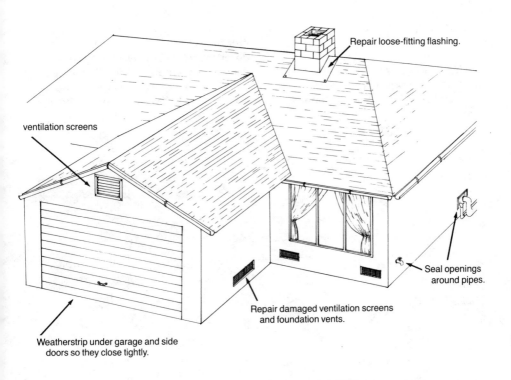

Repair loose-fitting flashing.

ventilation screens

Seal openings around pipes.

Repair damaged ventilation screens and foundation vents.

Weatherstrip under garage and side doors so they close tightly.

FIGURE VIII.1. Rats and mice can gain entry to buildings through a variety of openings. Rats require ½-inch opening; mice require only ¼ inch.

trap is commonly used for rats and mice; however, there are many other types of traps available. Multiple-capture mouse traps are more expensive but can be quite effective, especially when mouse trapping is done on a continual basis.

The kind of bait used for the trap is important. Nut meats, dried fruit, or bacon make excellent baits for rats and mice. The bait should be fastened securely to the trigger of the trap with light string, thread, or fine wire so the rodent will spring the trap in attempting to remove the food. Soft baits such as peanut butter can be used, but rats or mice frequently will eat soft baits without setting off the trap. Leaving traps baited but unset until the bait has been taken at least once improves trapping success by making the rodents more accustomed to the traps. Set traps so the trigger is sensitive and will spring easily.

The best places to set traps are in secluded areas where rats and mice are likely to travel and seek concealment. Droppings, gnawings, and damage indicate the presence of rodents, and areas where such evidence is found are usually good places to set traps. Place traps in natural travelways, such as along walls, so the rodents will pass directly over the trigger of the trap. (See fig. VIII.2.) Traps set along a wall should extend from the wall at right angles, with the trigger end nearly touching the wall. If traps are set parallel to the wall, they should be set in pairs to intercept rodents traveling from either direction. Traps can also be set on ledges or shelves or fastened with wire to branches, fences, pipes, or beams. (See fig. VIII.3.) Trapping in these elevated areas is extremely useful for controlling roof rats. In areas where children, pets, or birds might contact traps, use a box or barrier to keep them away.

Use as many traps as are practical so trapping time will be short and decisive. A dozen traps for a heavily infested home may be necessary, and 50 to 100 or more are not too many for larger buildings. Mice seldom venture far from their shelter and food

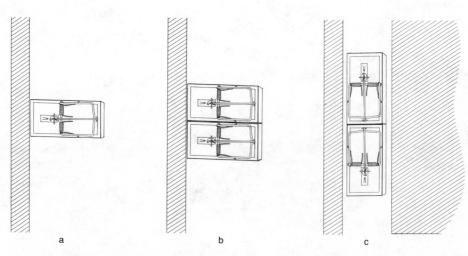

a b c

FIGURE VIII.2. Set traps along walls as shown (a, b), so rodent passes over trigger. In tight spaces, such as behind a refrigerator, two traps will have to be set (c).

supply, so place traps 10 feet or less apart in areas where signs of mice are observed. Rat traps can be placed somewhat farther apart.

Glue Boards

An alternative to traps is glue boards, especially for mice. These work on the same principle as flypaper: when a rat or mouse attempts to cross the glue board, the rodent gets stuck. Place glue boards in secluded areas along walls or in other places where rodents travel. If the rodent travelway includes a place where they jump from one level to another, place the glue board where it is likely to be jumped onto. It is not necessary to bait glue boards, although bait can entice rodents onto the board in some situations. Glue boards tend to lose their effectiveness in dusty areas, and extremes of temperature may affect the tackiness of the adhesive. In many cases, mice and rats trapped on glue boards will not die immediately. If this happens, kill them quickly by immersing the entire board in water.

FIGURE VIII.3. For roof rats, snap traps can be secured to tree limbs where rats run.

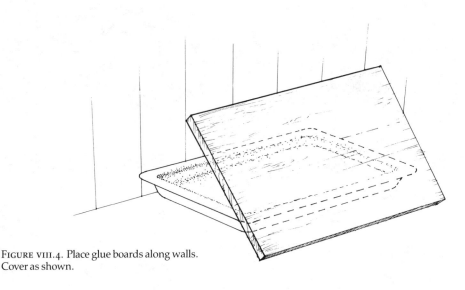

FIGURE VIII.4. Place glue boards along walls. Cover as shown.

Toxic Bait

When the number of rats or mice in or around a building is high, you may need to use toxic bait to achieve adequate control. Anticoagulant baits are probably the safest baits to use around the home and garden, but other types of bait can be used effectively. As you would with any poison, take care to ensure safety to children and pets. Additionally, poisoned rats and mice often die in inaccessible locations within a building, leading to persistent and unpleasant odors.

During the baiting process, dispose of dead rodents by burying them or placing them in plastic bags and putting them in the trash. Handle dead rodents with gloves. All bait containers should be clearly labeled with appropriate warnings. Store unused bait and containers in a locked cabinet inaccessible to children and domestic animals.

Baiting for house mice and Norway and roof rats. With house mice, proper placement of baits and the distance between placements is critical. Place baits close to where the mice are living and no farther than 10 feet apart (preferably closer).

Place all baits in rat travelways or near their burrows and harborages. Do not expect rats to go out of their way to find the bait. It is a good idea to place baits under cover so the rats will feel secure while feeding. Bait placements for Norway rats will generally not be effective for roof rats because they feed in different areas. For roof rats, place baits in elevated locations, such as in the crotch of a tree, on the top of a fence, or high in a vine. If you place bait above

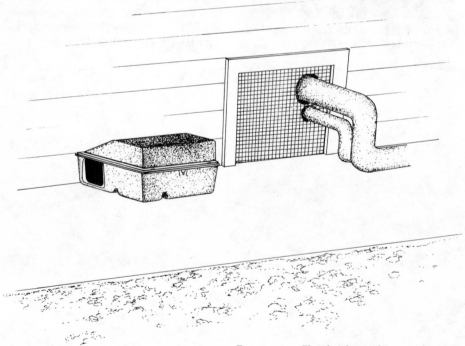

FIGURE VIII.5. Place bait boxes along travelways of Norway rats and house mice.

ground level, take care that the bait is securely fastened, so it will not fall to the ground where children or pets can find it. If necessary, cover the bait to prevent birds from eating it. (See figs. VIII.5 and 6.)

Rats must feed on most anticoagulant baits several times for the bait to be effective. For this reason, the bait must be available for 5 or more days. For best results, provide fresh bait each day until feeding stops. Anticoagulant baits are commonly available in meal, pellet, kernel, or liquid form. Paraffin blocks impregnated with toxic grains are available for use in areas where moisture might spoil other types of baits. (See fig. VIII.7.) Anticoagulants mixed with water and sugar are particularly useful when water is scarce or when food availability cannot be restricted.

Bait boxes. Bait boxes are used with anticoagulant baits. They protect the bait from the weather and restrict accessibility mainly to rodents, providing a safeguard for people, pets, and other animals. (See fig. VIII.8.) Bait boxes or other types of enclosed bait stations should be large enough to accommodate several rats or mice at a time and should contain a self-feeding hopper for holding the bait. Each station should have at least two 1-inch openings for mice and two 2 ½-inch openings for rats. Bait stations should be placed next to walls or in places where rats will encounter them. Commercial bait boxes are available in a variety of sizes and shapes. They can also be constructed to fit the situation, often from discarded materials. Water-resistant cardboard, wood, plastic, and metal have all been used. All bait boxes should be clearly

FIGURE VIII.6. For roof rats, bait stations can be secured in a tree.

FIGURE VIII.7. Roof rats will readily feed from anticoagulant paraffin bait blocks placed in trees. Be sure to secure them well so they don't fall and present a hazard to children or pets.

labeled *poison* and should be locked or secured.

Water baits. Rats and mice will drink water daily if it is available. Where access to water is restricted, anticoagulant water baits can be very effective. Usually, water baits are sweetened with up to 5 percent sugar to increase their attractiveness. Because water is consumed by most animals, it is not selective in its attractiveness; therefore, water baits should be used only where nontarget animals cannot gain access to them.

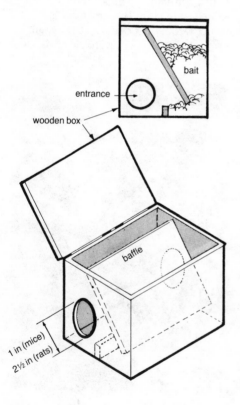

FIGURE VIII.8. Example of a bait box for rats and mice. Commercial boxes are also available.

Paraffin bait blocks. Bait blocks are generally used for baiting in moist areas. They have been found particularly valuable in situations where the blocks can be fastened near places roof rats frequent such as telephone poles, fences, or the rafters of a garage. It is usually easier to make bait blocks readily accessible to roof rats than to use bait boxes. Dogs may chew on a bait block so, as with all baits, place them out of the reach of dogs and other domestic animals. Norway and roof rats usually accept loose-type baits better than paraffin baits, so it is advisable to use bait blocks only when loose cereal baits are not practical.

Single-feeding baits. Single-feeding baits—those lethal after one feeding—are useful in some situations, particularly when rat and mouse populations are high or are spread over large areas. Zinc phosphide is a common single-feeding bait used for rats and mice. Place the bait in runways or next to burrows where rats and mice will find it. This bait is rapid acting and you may find dead rodents within 12 hours of baiting. Dispose of all dead rodents by burying them or placing them in plastic bags and putting them in the trash. Always use gloves to avoid contact with rodents. Because zinc phosphide does not accumulate in the tissue of the rodent, predators or scavengers such as dogs and cats are not likely to be adversely affected by eating poisoned carcasses. However, they, as well as other animals and children, can be affected by eating the bait so it must be stored and used carefully.

Take care to avoid bait shyness, a condition where rats and mice have eaten only enough bait to make them sick. Once rats and mice become bait shy, they will not eat zinc phosphide bait for 6 months or more. This will usually not occur if bait is used according to the label directions. Do not repeat treatments with zinc phosphide bait more often than every 6 months. Zinc phosphide is a Restricted Use Material and you need a permit from the local County Agricultural Commissioner to use it. Use of the bait by home owners may be exempted from the permit requirement in some situations when small amounts of bait are purchased.

Other Control Methods for Rats and Mice

Rats and mice are wary animals, easily frightened by unfamiliar or strange noises. However, they can quickly become accustomed to sounds that are repeated regularly. The use of frightening sounds is not considered effective for controlling rats and mice in home and garden situations.

Rats and mice have an initial aversion to some odors and tastes, but no repellents have been found to solve a rat or mouse problem. There are no rat or mouse repellents registered for use in California.

As with all animals, natural constraints such as 1) predation, 2) disease, or 3) inadequate food and shelter limit rat and mouse populations. Because animal numbers will not increase indefinitely, one alternative is to do nothing, letting the population limit itself. Experience has shown, however, that, particularly around homes and gardens, the point at which rat and mouse populations level off naturally is intolerably high.

Predators, especially cats and owls, eat rats and mice. However, in most cases, these predators are unable to keep rodents below damaging levels.

Monitoring Guidelines

Because rats and mice are active throughout the year, periodically check the area for signs of their presence. If possible, inspect property adjacent to your garden or structure because rats and mice may soon invade from this direction. Experience has shown it is less time consuming to control rodents before their numbers get too high and, if you use poison baits, less material will be required if control is started early.

Tree Squirrels

Western gray squirrel

There are four species of tree squirrels found in California that sometimes cause damage around homes and gardens. They feed on green and ripe walnuts, almonds, oranges, avocados, apples, strawberries, tomatoes, and grain. They sometimes gnaw on telephone cables and may chew into wooden buildings or invade attics through knotholes and other openings. Tree squirrels may carry certain diseases that are transmissible to humans. (See p. 26.)

The Western grey squirrel (*Sciurus griseus*) is a native tree squirrel found from the Mexi-

can border north through the coast ranges and from the Tehachapi mountains north along the western slope of the Sierra Nevada. The Eastern or red fox squirrel (*S. niger*) is an introduced species established in city parks and adjacent areas in Fresno, San Diego, San Mateo, Santa Cruz, San Fernando, Sacramento, San Francisco, and the South Bay area, and in agricultural lands east of Ventura and Oxnard. The Douglas squirrel (*Tamiascirus douglasii*) is a native of the north central area and the Sierra Nevada. The Eastern grey squirrel (*S. carolinensis*) has been introduced from the east into Golden Gate Park in San Francisco and is established in small areas of Calaveras and San Joaquin counties.

All of these tree squirrels are active during the day and are frequently seen in trees, on telephone lines, and foraging on the ground. Tree squirrels are easily distinguished from ground squirrels and chipmunks by their long, bushy tails and lack of dorsal spots or stripes. Although they are chiefly aboreal, some, particularly the red fox and western grey squirrels, spend considerable time foraging on the ground. Tree squirrels do not hibernate, but are active year-round except in inclement or very cold weather. They are most active in the early morning and late afternoon.

Legal Restraints of Control

The tree squirrels *S. griseus*, *S. carolinensis*, and *T. douglasii* are classified as game mammals by the California Fish and Game Code and can be taken only as provided by the hunting regulations. The red fox squirrel (*S. niger*) is an exception: when it is found injuring growing crops or other property, it may be taken (controlled) at any time or in any manner by the owner or tenant of the premises. Additionally, any owner or tenant of land or property that is being damaged or destroyed, or is in danger of being damaged or destroyed by grey squirrels, may apply to the Department of Fish and Game for a permit to control such animals. The Department, upon receipt of satisfactory evidence of such damage or destruction, actual or immediately threatened, will issue a revokable permit for the taking and disposition of such mammals under regulations promulgated by the Fish and Game Commission. These squirrels cannot be sold or shipped from the premises on which they were taken, except under instructions from the Department. No poison may be used on any grey squirrel and none is currently registered for the red fox squirrel. If a permit to trap the grey squirrel is issued, the Department will designate the type of trap to be used. The Department may also require squirrels to be released in parks or other nonagricultural areas.

Control Methods

Trapping

A good trapping program can eliminate the red fox squirrel from an area. A modified wooden box-type ground squirrel trap has been used quite successfully (see fig. IX.1). Instructions for making this trap can be found in the chapter entitled Ground Squirrels in this publication. A handful of nutmeats placed well behind the trigger mechanism will attract the squirrels. A few nuts

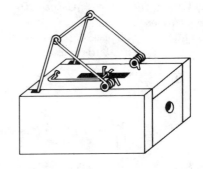

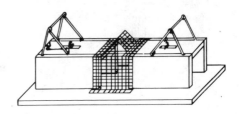

Figure IX.1. Box-type tree-squirrel trap

may be scattered at the trap entrance to entice the squirrels. For best results, leave baited traps unset for several days until the squirrels become accustomed to pushing back the swinging trigger loop to reach the bait. After the squirrels have become familiar with the traps, rebait and set all the triggers. Be sure to tie the bait to the trigger with fine thread or string.

If squirrels are entering the area via overhead routes such as trees or power lines, place traps on rooftops or secure traps to tree limbs (see fig. IX.2). Live-catch traps can also be used for capturing red fox squirrels; however, they present the problem of what to do with the live animal.

Ordinary wooden snap traps used for rats may be effective in controlling red fox squirrels. As with the box traps, prebaiting probably will improve their effectiveness. All bait shoud be tied to the trigger with thread or light string. These traps can be nailed or fastened to fences, tree limbs, or rooftops. Be careful of how you place traps in these areas—you might trap birds unless the traps are placed in such a way that birds cannot have easy access or entry to them.

A considerable number of red fox squirrels can be taken with few traps if the traps are kept in continuous operation. Trapping should be commenced as soon as damage is threatened or observed.

Other Control Methods

Exclusion (rodent-proofing) is the best solution to solving problems of tree squirrels gaining access to dwellings. Most entry points will be above eye level but exceptions do occur. Sheet metal or wire hardware cloth are most often used to close openings. When closing entry routes be sure you do not trap animals inside the building.

Although some repellents are registered for use on tree squirrels, their effectiveness remains in question.

As with all animals, natural population constraints limit tree squirrel numbers. Predators do eat tree squirrels; however, in most cases, they are unable to keep the squirrel populations below damaging levels.

Monitoring Guidelines

The important first step of detecting the presence of tree squirrels is fairly easy because they are active during daylight hours. If tree squirrels are observed in the garden area, it is likely that damage will occur at certain stages of crop development, particularly with nut crops. If tree squirrels are frequently seen in nut-bearing trees, some kind of preventive action should be taken; squirrels can strip the tree of nuts in a short time.

FIGURE IX.2. Box traps can be placed on tree limbs for tree squirrels.

Glossary

Acute rodenticide—Toxic compound specifically formulated to kill rodents from a single feeding.

Bait box—Small structure in which anticoagulant bait is placed so that the animal must enter to take the bait.

Browsing—To feed on leaves and shoots of plants and trees.

Burrow—An underground excavation used for shelter, food storage and rearing of young.

Communal roost—Roost used in common by a large number of birds.

Control—To regulate, restrain, or curb the population of a wildlife species that has become a pest.

Debarking—Removal of bark from a tree or woody plant by an animal.

Depredation—When a pest animal or species despoils or "plunders" a crop.

Diurnal—Animal or species of animal active during daylight hours.

Estivation—State of dormancy in an animal's activity cycle, occurring during summer months.

Exclusion—To prevent one or more animals or species of animals from entering an area.

Feeding flock—Flock of birds banded together for the purpose of feeding.

Flyway—Aerial path habitually used by a species of birds to travel from one point to another.

Food cache—A quantity of food stored by the animal for later use.

Forbs—Any small broad-leafed flowering plant.

Frightening device—Any device used to drive a pest species away from a crop area or other area by scaring it. Examples of frightening devices are gas cannons, shell crackers, flags, and noisemakers.

Fumigant—Substance that produces toxic or suffocating gases.

Game mammal—Mammal specified by the California Fish and Game Code to be hunted for food or sport.

Girdle—Gnawing cut made by an animal encircling the trunk or limb bark of a tree or other woody plant.

Habitat modification—To alter the environment where an animal species is found. In wildlife pest control this is sometimes done to make the habitat less favorable for the species concerned.

Hibernation—State of dormancy in an animal's activity cycle occurring during winter months.

Introduced species—A species released in an area where it was previously not found.

Leg-hold trap—Device for trapping and holding or restraining an animal by the leg.

Lethal dose—Quantity of a toxicant necessary to cause death.

Migratory bird—A bird that makes an annual, regular round trip between two geographic regions.

Multiple-dose bait—Poisonous bait that requires a sustained dosage over a period of time to produce death. An example is an anticoagulant.

Native bird—A bird that is an original species in an area or region. Indigenous.

Natural control—Regulation or restraint of an animal population in a manner conforming to nature. Examples: predators, diseases, lack of food or cover.

Netting—Nets used to protect a crop from birds.

Nongame mammal—Any animal not commonly hunted, as specified in the California Fish and Game Code.

Nontarget species—Any species that is not the object of the control being applied.

Phytotoxic—Injurious and sometimes lethal to plants.

Prebaiting—Placing of nontoxic bait to condition a pest species to eating it before toxic bait is applied or so the pest will become accustomed to a trap.

Predator—Any animal that survives by regularly taking another animal (or insect) for food.

Resident species—Species that lives in an area year-round. Does not migrate.

Runway—Path that an animal or animals commonly travel over or through.

Single-feeding (dose) bait—Toxic bait that produces death from one dose. Also called acute toxic bait.

Spot baiting—Placing of bait by hand at selected sites.

Untreated bait—Bait to which no toxic or repellent substance has been applied. Used for prebaiting.

Wildlife pest—Any species of wild animal, in any area, that becomes a health hazard, causes economic damage, or is a general nuisance to one or more persons.